Contents

Mathematical Modeling of Biofilms

Scientific and Technical Report No.18

Mathematical Modeling of Biofilms

IWA Task Group on Biofilm Modeling:
Hermann Eberl, Eberhard Morgenroth,
Daniel Noguera, Cristian Picioreanu,
Bruce Rittmann, Mark van Loosdrecht
and Oskar Wanner

Publishing

Published by IWA Publishing, Alliance House, 12 Caxton Street, London SW1H 0QS, UK

Telephone: +44 (0) 20 7654 5500; Fax: +44 (0) 20 7654 5555; Email: publications@iwap.co.uk
Web: www.iwapublishing.com

First published 2006
© 2006 IWA Publishing

Index prepared by Indexing Specialists, Hove, UK.
Printed by Lightning Source

British Library Cataloguing in Publication Data
A CIP catalogue record for this book is available from the British Library

Library of Congress Cataloging- in-Publication Data
A catalog record for this book is available from the Library of Congress

ISBN 1843390876
ISBN13: 9781843390879

List of Task Group members

Oskar Wanner
Urban Water Management Department, Swiss Federal Institute of Environmental Science and Technology (EAWAG), Switzerland

Hermann J. Eberl
Department of Mathematics and Statistics, University of Guelph, Canada

Eberhard Morgenroth
Department of Civil and Environmental Engineering and Department of Animal Sciences, University of Illinois at Urbana-Champaign, USA

Daniel R. Noguera
Department of Civil and Environmental Engineering, University of Wisconsin – Madison, USA

Cristian Picioreanu
Department of Biotechnology, Delft University of Technology, The Netherlands

Bruce E. Rittmann
Center for Environmental Biotechnology, Biodesign Institute at Arizona State University, USA

Mark C.M. van Loosdrecht
Department of Biotechnology, Delft University of Technology, The Netherlands

Acknowledgements

This report was prepared by the IWA Task Group on Biofilm Modeling. The following provided important assistance with the solution of the benchmark problems and preparation of several sub-sections of the report:
• Gonzalo E. Pizarro, formerly at the University of Wisconsin, Madison (USA) and now at the Universidad Católica de Chile
• Alex Schwarz, formerly at Northwestern University and now with BSA Consultores e Ingieneros, Santiago, Chile
• Julio Pérez, formerly at the Delft University of Technology and now at the Universidad Autonoma de Barcelona, Spain

Parts of this report were presented at the IWA Biofilm Specialists Conference in Cape Town, South Africa (September 2003) and have been published in modified form in *Water Science and Technology* Vol. 49, no. (11-12), 2004. Other parts of this report were presented at the IWA Biofilm Specialists Conference in Las Vegas, Nevada, USA (October 2004).

The Task Group greatly appreciates the financial support of IWA.

Overview

WHAT IS A BIOFILM?

The simple definition of a biofilm is "microorganisms attached to a surface." A more comprehensive definition is "a layer of prokaryotic and eukaryotic cells anchored to a substratum surface and embedded in an organic matrix of biological origin." Some biofilms are good, providing valuable services to human society or the functioning of natural ecosystems. Other biofilms are bad, causing serious health and economic problems. Understanding the mechanisms of biofilm formation, growth, and removal is the key for promoting good biofilms and reducing bad biofilms.

The two definitions at the previous paragraph underscore that a biofilm can be viewed simply or by taking into account complexities. The "better" definition depends on what we want to know about the biofilm and what it is doing. Mathematical modeling is one of the essential tools for gaining and applying this kind of mechanistic understanding of what the biofilm is and is doing.

WHAT IS A MODEL?

A mathematical model is a systematic attempt to translate the conceptual understanding of a real-world system into mathematical terms. A model is a valuable tool for testing our understanding of how a system works.

Creating and using a mathematical model require six steps.
1. The important variables and processes acting in the system are identified.
2. The processes are represented by mathematical expressions.

© IWA Publishing 2006. *Mathematical Modeling of Biofilms: Scientific and Technical Report No. 18* by the IWA Task Group on Biofilm Modeling (Hermann Eberl, Eberhard Morgenroth, Daniel Noguera, Cristian Picioreanu, Bruce Rittmann, Mark van Loosdrecht and Oskar Wanner). ISBN: 1843390876. Published by IWA Publishing, London, UK

3. The mathematical expressions are combined together appropriately in equations.
4. The parameters involved in the mathematical expressions are given values appropriate for the system being modeled.
5. The equations are solved by a technique that fits the complexity of the equations.
6. The model solution outputs properties of the system that are represented by the model's variables.

Modeling is a powerful tool for studying biofilm processes, as well as for understanding how to encourage good biofilms or discourage bad biofilms. A mathematical model is the perfect means to connect the different processes to each other and to weigh their relative contributions.

Mathematical models come in many forms that can range from very simple empirical correlations to sophisticated and computationally intensive algorithms that describe three-dimensional (3d) biofilm morphology. The best choice depends on the type of biofilm system studied, the objectives of the model user, and the modeling capability of the user.

Starting in the 1970s, several mathematical models were developed to link substrate flux into the biofilm to the fundamental mechanisms of substrate utilization and mass transport. The major goal of these first-generation mechanistic models was to describe mass flux into the biofilm and concentration profiles within the biofilm of one rate-limiting substrate. The models assumed the simplest possible geometry (a homogeneous "slab") and biomass distribution (uniform), but they captured the important phenomenon that the substrate concentration can decline significantly inside the biofilm.

Beginning in the 1980s, mathematical models began to include different types of microorganisms and non-uniform distribution of the biomass types inside the biofilm. These second-generation models still maintained a simplified 1-dimensional (1d) geometry, but spatial patterns for several substrates and different types of biomass were added. A main motivation for these models was to evaluate the overall flux of substrates and metabolic products through the biofilm surface.

Starting in the 1990s and carrying to today, new mathematical models are being developed to provide mechanistic representations for the factors controlling the formation of complex 2- and 3-d biofilm morphologies. Features included in these third-generation mathematical models usually are motivated by observations made with the powerful new tools for observing biofilms in experimental systems.

Today, all of the model types are available to someone interested in incorporating mathematical modeling into a program of biofilm research or application. Which model type to choose is an important decision. The third-generation models can produce highly detailed and complex descriptions of biofilm geometry and ecology; however, they are computationally intense and demand a high level of modeling expertise. The first-generation models, on the other hand, can be implemented quickly and easily – often with a simple spreadsheet – but cannot capture all the details. The "best" choice depends on the intersection of the user's modeling capability, biofilm system, and modeling goal.

MODEL SELECTION

The first step in creating or choosing a biofilm model is to identify the essential features of the biofilm system. Features are organized into a logical hierarchy that is illustrated in Figure 0.1:

- Compartments define the different sections of the biofilm system. For example, the biofilm itself is distinguished from the overlying water and the substratum to which it is attached. A mass-transport boundary layer often separates the biofilm from the overlying water.

- Within each compartment are <u>components</u>, which can include the different types of biomass, substrates, products, and any other material that is important to the model. The biomass is often divided into one or more active microbial species, inert cells, and extracellular polymeric substances (EPS).
- The components can undergo <u>transformation</u>, <u>transport</u>, and <u>transfer</u> <u>processes</u>. For example, substrate is consumed, and this leads to the synthesis of new active biomass. Also, active biomass decays to produce inerts.
- All processes affecting each component in each compartment are mathematically linked together into a <u>mass balance equation</u> that contains <u>rate terms</u> and <u>parameters</u> for each process.

Because most biofilms are complex systems, a biofilm model that attempts to capture all the complexity would need to include (i) mass balance equations for all processes occurring for all components in all compartments, (ii) continuity and momentum equations for the fluid in all compartments, and (iii) defined conditions for all variables at all system boundaries. Implementing such a model is impractical, maybe impossible. Therefore, even the most complex biofilm models existing today contain many simplifying assumptions. Most biofilm models today capture only a small fraction of the total complexity of a biofilm system, but they are highly useful. Thus, simplifications are necessary and a natural part of modeling. In fact, the "golden rule" of modeling is that *a model should be as simple as possible, and only as complex as needed.*

Good simplifying assumptions are identified by a careful analysis of the characteristics of a specific system. These good assumptions become part of the model structure; in other words, they serve as guidance for the selection of the model. The models found in the literature can be differentiated by their assumptions, which depend on the objectives of the modeling effort and the desired type of modeling output. Thus, a user that is searching for a model to simulate specific features of a biofilm system should begin by evaluating the type of assumptions used in creating the models.

One of the objectives of the IWA Task Group on Biofilm Modeling was a comparison of characteristic biofilm models using benchmark problems. A main purpose was to analyze the significance of simplifying assumptions as a prelude for providing guidance on how to select a model.

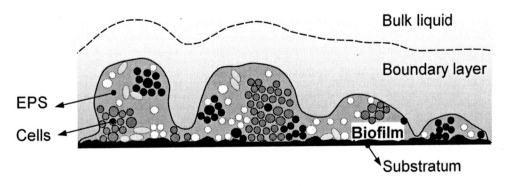

Figure 0.1. Four compartments typically defined in a biofilm system: bulk liquid, boundary layer, biofilm and substratum.

Table 0.1. Features by which various types of models of biofilm systems differ. Model codes are: (A) analytical, (PA) pseudo-analytical, (N1) 1d numerical, and (N2/N3) 2d/3d numerical.

Feature	A	PA	N1	N2/N3
Development over time (i.e., dynamic)	-	-	+	+
Heterogeneous biofilm structure	-	-	o	+
Multiple substrates	o	o	+	+
Multiple microbial species	o	o	+	+
External mass transfer limitation predicted	o	o	+	+
Hydrodynamics computed	-	-	-	+

The models used by the Task Group can be grouped into four distinct categories according to the level of simplifying assumptions used: namely, analytical (A), pseudo-analytical (PA), 1d numerical (N1), and 2d/3d numerical (N2/N3). As a baseline, all model types normally can represent biofilms having the following features: (i) the biofilm compartment is homogeneous, with fixed thickness and attached to an impermeable flat surface, (ii) only one substrate limits the growth kinetics, (iii) only one microbial species is active, (iv) the bulk liquid compartment is completely mixed, and (v) the external resistance to mass transfer of dissolved components is represented with a boundary layer compartment with a fixed thickness.

Table 0.1 identifies other features that can be incorporated into certain models and that differentiate among the model types. A plus sign (+) means that the feature can be simulated, a minus sign (-) indicates that the model cannot simulate that feature, and a zero (o) indicates that the model may be able to simulate the feature, but with restrictions. In general, the flexibility and complexity of the models is lower on the left hand side of the table and increases towards the right hand side.

Selecting a model is intimately related to the modeling objectives and the modeling capability of the user of the model. Common quantitative objectives are the calculation of substrate removal, biomass production and detachment rates, or the quantity of biomass present in a given biofilm system. In engineering applications, biofilm models also are employed to optimize the operation of existing biofilm reactors and to design new reactors. In research, they serve as tools to fill gaps in our knowledge, as they help to identify unknown processes and to provide insight into the mechanisms of these processes. The capability of the user relates to the computing power available and, equally important, to the user's capacity for understanding the model. A model that cannot be formulated or solved by the user is of no value, whether or not it addresses the objectives well.

Biofilm models can be used to provide information at macro-scale or micro-scale. *Macro-scale* outputs include substrate removal rates, biomass accumulation in the biofilm and biomass loss from the system. Typical *micro-scale* outputs are the spatial distributions of substrates and microbial species in the biofilm.

Simplifying assumptions are related to making the modeling objective mesh with the user capability. For instance, if the objective is to describe the performance of a biofilm system at the macroscale, then the various compartments and processes do not need to be described in too much of a microscale. A lot of microscale detail makes the model difficult to create and computing-intensive. For example, a 1d model with only one type of active biomass may be completely adequate to estimate the flux of one substrate averaged over square meters.

If the objective is to model micro-scale processes (e.g., the interaction between microbial cells and EPS in the biofilm or 3d physical structures at the µm-scale), the number and type of processes occurring in each compartment of the biofilm need to be represented in microscale detail. For example, a 2d or 3d model is necessary if understanding the physical structure of the biofilm at the µm-scale is the modeling objective, while a multi-species model is necessary if the objective is to understand how ecological diversity develops. When microscale detail is required, the size of the system being modeled will need to be small in order to make the model's solution possible.

Although many processes always take place in a biofilm, it is not necessary to include every one, depending on the objectives. For example, the spatial distribution of the particulate components can be specified by an *a priori* assumption, instead of predicted by the model, if the goal is to predict substrate flux for a known biofilm. Then, the model needs not include the processes of microbial growth and loss. On the other hand, when the objective is to predict the distribution of microbial species within the biofilm or to calculate the expected biofilm thickness at steady state, then microbial growth and detachment processes are essential.

BIOFILM MODELS

The most basic principle for all quantitative models is conservation of mass. Conservation of mass of a component in a dynamic and open system states that:

$$
\begin{pmatrix} \textbf{\textit{Net rate of}} \\ \textbf{\textit{accumulation}} \\ \textit{of mass} \\ \textit{of component} \\ \textit{in the system} \end{pmatrix} = \begin{pmatrix} \textit{Mass flow} \\ \textit{of the} \\ \textit{component} \\ \textbf{\textit{into}} \\ \textit{the system} \end{pmatrix} - \begin{pmatrix} \textit{Mass flow} \\ \textit{of the} \\ \textit{component} \\ \textbf{\textit{out of}} \\ \textit{the system} \end{pmatrix} + \begin{pmatrix} \textit{Rate of} \\ \textbf{\textit{production}} \\ \textit{of the} \\ \textit{component by} \\ \textit{transformations} \end{pmatrix} - \begin{pmatrix} \textit{Rate of} \\ \textbf{\textit{consumption}} \\ \textit{of the} \\ \textit{component by} \\ \textit{transformations} \end{pmatrix}
$$

The local mass balances are the mathematical form of equality, which in a Cartesian space (i.e., with ortho-normal unit vectors) can be written as

$$
\frac{\partial C}{\partial t} = -\frac{\partial j_x}{\partial x} - \frac{\partial j_y}{\partial y} - \frac{\partial j_z}{\partial z} + r
$$

where t is time (T); x, y and z are spatial coordinates (L); C is the concentration (ML^{-3}); j_x, j_y and j_z, are the components of the mass flux $\mathbf{j}$ (ML^{-2}T^{-1}) along the coordinates; and r is the net production rate (ML^{-3}T^{-1}) of the component. This is the *equation of continuity* for a component, either soluble or particulate.

At the macroscopic level, *global* mass balances can be written based on the continuity over the whole biofilm system. The global mass balances result also from integration of the local balances and constitute the main engineering form of mass balance. The global mass balances state that, for any dissolved or particulate component, the change of component mass in time in the system is equal to the difference between component mass flow rate in influent and effluent, plus the net production rate in the system volume. In mathematical terms, for any component, this is written as

$$
\frac{dm}{dt} = F_{in} - F_{ef} + F_{gen}
$$

where m is the component mass (M), F_{in} and F_{ef} are the component mass flow rates in the influent and the effluent (MT^{-1}), respectively, and F_{gen} is the sum of the rates of all the processes by which the component is produced or consumed (MT^{-1}). If two compartments, a completely mixed "bulk liquid" and "biofilm", are distinguished in the system, the equation of continuity for the bulk liquid compartment becomes

$$\frac{d\left(V_B C_B\right)}{dt} = F_{in} - F_{ef} + F_F + F_B$$

where F_B and F_F are the overall transformation flow rates in the bulk liquid and biofilm (MT^{-1}), respectively, C_B is bulk liquid (and effluent, too) component concentration (ML^{-3}), and finally, V_B is volume of bulk liquid phase (L^3).

All models analyzed in this report derive from the same general principle of mass conservation for soluble and particulate components. The models differ, however, by the number and the level of simplifying assumptions made to provide a solution. The most general biofilm description would describe the development in time of a 3d distribution of multiple soluble and multiple particulate components under diverse hydrodynamic conditions. Due to the complexity of the mathematical description, such a model requires a numerical solution – in fact a very sophisticated numerical solution that demands a very high-capacity computer. However, such complex and comprehensive model is not always necessary. The benchmark problems are good examples of settings for which much simpler models can work well.

An *analytical* (A) model is the simplest solution of the general biofilm reactor model. The determinative feature of an A model is that its solution is obtained by mathematical derivation and without any numerical techniques. Advantages of an A solution, in addition to its being in a simple equation format, is that the effects of each term, variable, or parameter (e.g., diffusion coefficient, microbial kinetics, and substrate concentration) can be directly analyzed. The disadvantage of an A solution is that the biofilm system must be very simple to yield a mathematically derivable solution. Multiple components, complex geometries, and time dynamics are difficult, if not impossible, to include and still have an A solution.

Analytical models are most useful for evaluating biofilm systems that have one dominant process (e.g., nitrification or BOD removal). An A model also can be applied for multi-species + multi-substrate systems when significant *a priori* knowledge of biofilm composition is available. Analytical models are not well suited for predicting the exact distributions of different types of bacteria in the biofilm, the conversion of multiple substrates, the total biofilm accumulation, or complex biofilm structure.

A *pseudo-analytical* (PA) model is a simple alternative when one or more of the simplifications used in an A model must be eliminated to gain a realistic representation of the biofilm system. PA solutions are comprised of a small set of algebraic equations that can by solved directly by hand or with a spreadsheet. The solution outputs the substrate flux (J) when the bulk-liquid substrate concentration (S) is input to it. The relative ease of using the PA solutions makes them amenable for routine application in process design and as a teaching tool. The pseudo-analytical solution is simply coupled with a reactor mass balance so that the unique combination of substrate concentration and substrate flux in the reactor is computed for a given biofilm system.

The PA solution for a steady-state biofilm was developed for single-substrate and single-species setting, but can be applied for multi-species biofilms. Such PA solutions make multi-species modeling more accessible to students, engineers, and non-specialist researchers. In addition, creating and using a multi-species PA model illuminates the important interactions that take place among the different types of biomass in a multi-species biofilm.

The *numerical 1d* (N1) models represent multi-species and multi-substrate biofilms in one dimension perpendicular to the substratum. Their complexity lies between the simpler A and PA models and the numerically demanding multi-d models. The N1 model equations must be solved numerically, but even complex simulations can be performed on a PC within minutes. The most significant feature of an N1 model is its flexibility with regard to the number of dissolved and particulate components, the microbial kinetics, and to a certain extent also the physical and geometrical properties of the biofilm. An N1 model can be used as a tool in research, as well as for the design and simulation of biofilm reactor. Already available commercial simulation software that implements such N1 models make also dynamic multi-species modeling accessible to students, engineers, and non-specialist researchers.

Examples of particulate components are active microbial species, organic and inorganic particles, and EPS. Examples of dissolved components are organic and inorganic substrates, metabolites, products, and the hydrogen ion. The output produced by the N1 model includes

- Spatial profiles of any number of particulate components in the biofilm
- Accumulation and the loss from the system of the mass of the particulate components
- Spatial profiles of any number of dissolved components in the biofilm
- Removal rates and effluent concentrations of the dissolved components
- Biofilm thickness as a function of the production and decay of particulate material in the biofilm and of attachment and detachment of cells and particles at the biofilm surface and in the biofilm interior

For all these quantities, the development in time, as well as steady state solutions, can be calculated.

Numerical 2d and 3d (N2 and N3) models are used to describe the heterogeneous characteristics of biofilms. The premise is that, by capturing the spatial and temporal heterogeneity of the physical, chemical, and biological environment, the model makes it possible to assess biofilm activity and interactions at the microscale. New problems to be addressed by multi-dimensional biofilm models include, for example:

- *Geometrical structure of biofilms*: How does the spatial biofilm structure form? What is the influence of environmental conditions on the biofilm structure? How does quorum sensing operate? What causes biomass detachment? How does microbial motility influence biofilm formation?
- *Mass transfer and hydrodynamics in biofilms*: What is the importance of advective mass transport relative to diffusion in the biofilm? How does the biofilm's spatial structure affect the overall solute transport rates to/from the biofilm?
- *Microbial distribution in biofilms*: What is the importance of inter-species substrate transfer? What is the influence of substrate gradients on microbial competition and selection processes?

The main difference between N1 and N2/N3 models is in the way processes affecting the development of the solid biofilm matrix and the dynamics of its composition (i.e., biomass growth, decay, detachment and attachment) are represented. For example, when second or third dimensions are part of the physical domain being modeled, the biofilm matrix has more than one direction in which to grow, allowing the simulation of spatially heterogeneous biofilms. Other potentially important phenomena that can be included with a multi-d model are fluid motion and advective mass transport in and out of the biofilm. Another class of addressed problems concerns the interaction among biofilm shape, fluid flow, biomass decay, and biofilm detachment.

Of course, allowing more complexity in the model increases the computing requirements dramatically. However, although initially some of the N2 and N3 models were coded and run using high performance supercomputers, nowadays, most multi-d biofilm models can be executed on single-processor machines.

BENCHMARK PROBLEMS

The benchmark problems help identify the trade offs inherent to using the different types of models. Because each model solves the same benchmark problem, the differences of the output produced by the models reflect the differences of the model complexity and of the simplifying assumptions that are made in the various models.

The first benchmark problem (BM1) describes nearly the simplest system possible: a mono-species biofilm that is flat and microbiologically homogeneous. Benchmark problem 2 (BM2) evaluates the influence of hydrodynamics on substrate mass transfer and conversions in a geometrically heterogeneous biofilm. Benchmark problem 3 (BM3) describes competition between different types of biomass in a multi-species and multi-substrate biofilm. Because the benchmark problems were designed to evaluate the ability of the models to represent fundamental features of a biofilm system, the trends apply to biofilms in treatment technology, nature, and situations in which biofilm is unwanted.

• BM1 describes a simple flat, mono-species biofilm. BM1 gives a baseline comparison of the different biofilm models for a biofilm system that is well suited for any modeling approach. The specific objective of BM1 is to compare key outputs, particularly including effluent substrate concentrations and substrate flux. Furthermore, the user friendliness of the different modeling approaches is evaluated.

For the simple conditions of BM1, modeling results are not significantly different for all modeling approaches having flat biofilm morphology. On the other hand, modeling results are strongly influenced by the assumption for mass transfer in the pores within the biofilm when heterogeneous biofilm morphology is allowed.

While modeling results are similar for most modeling approaches, the effort in implementing and using the different models is not. A and PA models can be readily solved using a spreadsheet. However, A or PA solutions require a number of simplifications, and the modeler has to make *a priori* decisions, e.g., on the dissolved component that is rate limiting. N1 models can be solved readily on a PC using available software. N2 and N3 models are able to simulate heterogeneous biofilm morphology, but they require custom-made software and, in some situations, extensive computing power. To approximately evaluate the influence of a heterogeneous morphology, N1 simulations can be combined to create pseudo-N3 models.

Thus, for simple biofilm systems and more-or-less smooth biofilm surfaces, A, PA, or N1 models often provide good compromise between the required accuracy of modeling results and the effort involved in producing these results. Adopting the more complex and intensive N2 and N3 models is justified only when the heterogeneity that they allow is critical to the modeling objective.

• BM2 involves spatially heterogeneous architectures that can induce complex flow patterns and affect mass transport. Classical 1d biofilm models are not able to capture this kind of complexity, which historically has been one of the reasons for the development of multi-d models.

Specifically, the assumption of a completely mixed bulk fluid is given up in this benchmark problem, and mass transport due to diffusion and advection in the fluid compartment are explicitly considered. The latter implies that the hydrodynamic flow field should be taken into account as well. A direct micro-scale mathematical description leads to a non-linear system of 3d partial differential equations in a complicated domain, and this is numerically expensive and difficult to solve. Therefore, BM2 investigates to what extent such a detailed local description of physical and spatial effects is necessary for macro-scale applications, where the purpose of the modeling is often only to calculate the total mass fluxes into the biofilm, *i.e.,* the global mass conversion rates. To this end, the description of the physical complexity of the system, expressed in geometrical and hydrodynamic complexity, can be simplified in various ways and to varying degrees in order to obtain faster simulation methods. The results of these simplified models are compared with the results for the fully 3d simulations.

Due to the high computational demand of 3d fluid dynamics in irregular domains, the problem is restricted to a small section of a biofilm (1.6 mm long). The goal of BM2 is to calculate: (1) the flux of dissolved substances into the solid region and, (2) the average substrate concentrations at the solid/liquid interface and at the substratum. A crucial aspect of the formulation of BM2 is the specification of appropriate boundary conditions. These are required to connect the small computational system with the external world that surrounds the modelling domain.

The 1d, 2d, and 3d models considered showed the same general sensitivities towards changes in biofilm thickness and hydrodynamics and were able to describe the qualitative system behavior. For the quantitative details, the key to a successful model reduction is a good description of the hydrodynamic conditions in the reactor segment. In a 2d reduction, this can be accomplished by a 2d version of the governing flow equations. The 1d approaches require a global mass balance or an empirical correlation that incorporates the hydrodynamics with passable reliability and accuracy.

Which simplified predictive model offers the best effort/accuracy/reliability trade-off depends largely on the hydrodynamic regime. Therefore, an analysis of the flow conditions in the reactor is required first before a simplification should be applied. Due to the enormous requirement of input data, the application of 3d models including full hydrodynamic calculations is restricted. These statements are made for applications in which only global results are of interest; that is, no refined resolution of the processes inside the biofilm is required. If such local results are desired, 1d models cannot yield a good description for spatially heterogeneous biofilms. At least a 2d model must be applied and, hence, the required input data and computing power must be provided.

- The goal of multi-species BM3 is to evaluate the ability of the different biofilm models to describe microbiological competition. In particular, BM3 focuses on competition for the same substrate and the same space in a biofilm. Together, these competitions provide a rigorous test for modeling multi-species biofilms, but without introducing unnecessary complexity.

To meet the goal, BM3 includes three biomass types having distinctly different metabolic functions:
- aerobic heterotrophs
- aerobic, autotrophic nitrifiers
- inert (or inactive) biomass

This scenario represents a common situation for biofilms in nature and in treatment processes for wastewater and drinking water.

For simplicity and comparability, BM3 uses the same physical domain as BM1: a flat biofilm substratum in contact with a completely mixed reactor experiencing a steady flow rate. To avoid unnecessary complexity, BM3 treats the nitrifiers as one "species" that oxidizes NH_4^+-N directly to NO_3^--N. Thus, it does not consider the intermediate NO_2^- or the division of nitrifiers between ammonia oxidizers and nitrite oxidizers. Active heterotrophs and nitrifiers follow Monod kinetics for substrate utilization and synthesis. They also undergo decay following two paths: (1) lysis and oxidation by endogenous respiration, and (2) inactivation to form inert or inactive biomass. The inert biomass does not consume substrate, and it is not consumed by any reactions. All forms of biomass can be lost by physical detachment.

The results of BM3 demonstrate that a wide range of 1d models is capable of representing the important interactions that can occur in biofilms in which distinctly different types of biomass can co-exist. The choice of the model depends on the user's needs and the modeling situation. One key choice is between models that demand a full numerical solution versus those that can be implemented with a spreadsheet. A second choice concerns the way in which the biomass is distributed. By far the simplest approach is to assume that the biomass types are independent of each other. This approach may work well when protection of a slow-growing species (like the nitrifiers) or dilution of a fast-growing species (like the heterotrophs) is not a major issue. When protection of a slow-growing species is critical to an accurate representation, then a model that accumulates the slow growers away from the outer surface is essential. When the dilution of a fast-growing species by slower growers is key, then a model that distributes the different biomass types throughout the biofilm is essential.

BM3 also was solved with two N2 models, which produce results similar to two N1 models for bulk substrate concentrations and fluxes into the biofilm. The similarity in output parameters for substrate concentration and fluxes is likely the result of the system in BM3 being a flat biofilm in a completely mixed reactor. On the other hand, the predicted distributions of the different types of biomass varied considerably between N1 and N2 models. In this regard, the only identifiable trend when comparing the N1 to the N2 models is the apparent increased, stable protection of nitrifiers in the N1 models, especially when this population is affected by a large accumulation of inerts or a lower rate of oxygen utilization. This trend likely reflects the different mechanisms to distribute the biomass within the biofilm.

1

Introduction

1.1 WHAT IS A BIOFILM?

The simple definition of a biofilm is "microorganisms attached to a surface." A more comprehensive definition is "a layer of prokaryotic and eukaryotic cells anchored to a substratum surface and embedded in an organic matrix of biological origin" (Wilderer and Characklis 1989). The importance of biofilms has steadily emerged since their first scientific description in 1936 (Zobell and Anderson 1936) and the first recognition of their ubiquity in the 1970s (Marshall 1976; Costerton *et al.* 1978). It is now estimated that planktonic microorganisms constitute less than 0.1% of the total aquatic microbial life (Costerton *et al.* 1995); thus, biofilms seem to constitute the preferred form of microbial life.

The ability of bacteria to attach to surfaces and to form biofilms can become an important competitive advantage over bacteria growing in suspension. Bacteria in suspension can be washed away with the water flow, but bacteria in biofilms are protected from washout and can grow in locations where their food supply remains abundant. The physical structure of the biofilm also allows for distinct biological *niches* that facilitate the growth and survival of microorganisms that could not compete successfully in a completely homogeneous system. Furthermore, microbial activity in biofilms can modify the internal environment (e.g., pH, O_2, metabolic products, or disinfectant concentration) to make the biofilm more hospitable than the bulk liquid (Rittmann and McCarty 2001).

Within a biofilm, a variety of microbial groups can contribute to the conversion of different organic and inorganic substrates. For example, when a wastewater contains a mixture of conventional and xenobiotic organic pollutants, biodegradation of the xenobiotics requires a population of slow-growing organisms – those capable of degrading the xenobiotics. The slow growers could be washed out of a suspended-growth process, since all the biomass has the same growth rate, which normally is controlled for the benefit of the bacteria that degrade the conventional organic pollutants. However, in a biofilm, the slow-growing bacteria can establish themselves deeper inside the biofilm, protected from loss, while the conventional pollutants are removed near the biofilm-fluid interface (Rittmann *et al.* 2000). The same situation can occur for slow-growing microorganisms that are undesirable, such as enteric pathogens that do not grow well in environmental conditions.

1.2   GOOD AND BAD BIOFILMS

Some biofilms are good, providing valuable services to human society or the functioning of natural ecosystems. Other biofilms are bad, causing serious health and economic problems. Figure 1.1 illustrates a number of good and bad biofilms, which are discussed briefly in this section.

Biofilms have been used to treat wastewater since the end of the 19th century. For example, the first trickling filter (Figure 1.1a) was placed in operation in England in 1893. Wastewater flowed into a basin filled with broken stones (the substratum) from the top and trickled down over the stones. A biofilm grew attached to the rocks, which provided a specific surface area of about 40 m^2/m^3 (Tchobanoglous *et al.* 2003). Trickling filters remain in common use today using either rock media or plastic media. The latter came into play in the 1970s and increased the substratum's surface area to about 200 m^2/m^3.

Rotating biological contactors (RBCs) (Figure 1.1d), also developed in the 1970s, have plastic biofilm media attached to a rotating axle. The media are partially submerged in the wastewater and continuously rotate, providing intermittent contact of the biofilm with the wastewater and atmospheric oxygen.

Low maintenance and stable operation are advantages of trickling filters and rotating biological contactors. However, the volumetric conversion rates for these systems are relatively low. Biofilm reactors with larger specific surface areas were developed starting in the 1980s. Biological filters (Figure 1.1b,c) for the treatment of wastewater and water use gravel-sized granular medium with specific surface areas up to 1000 m^2/m^3, which allow higher volumetric conversion rates. To prevent clogging, these biological filters have to be backwashed. Fluidized bed reactors (Figure 1.1f) are operated with increased upflow water velocities that suspend small carrier particles in the water phase. In airlift reactors (Figure 1.1e), the carrier particles are suspended in the circulating water flow that is caused by the injection of air. Specific surface areas from 2000 to 4000 m^2/m^3 can be achieved with the small carrier particles used in fluidized bed or airlift reactors. Added advantages can include better control of the biofilm due to uniform shear on the biofilm particles and no problems with liquid or gas distribution in the bottom of the reactor. In the moving bed biofilm reactor (Figure 1.1g), the carrier material has a density similar to the density of water. As a result, even larger particles (> 5 mm) can be suspended using a mixer or by aerating the reactor to create airlift pumping. With suspended support media, the moving bed biofilm reactor does not have to be backwashed as biofilm detachment is caused by particle-particle collisions within the system, but using larger carrier material results only in moderate specific surface areas of 330 m^2/m^3 (Rusten *et al.* 2000).

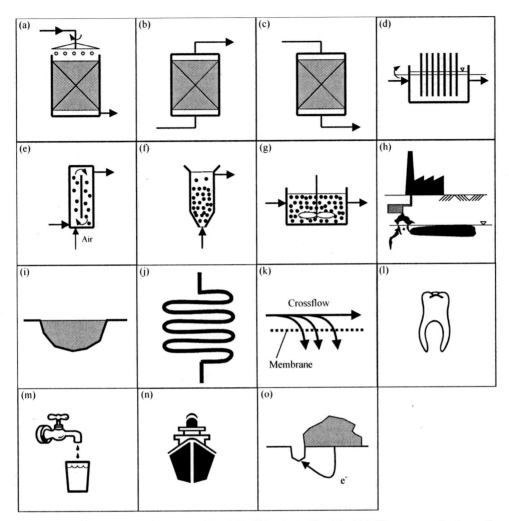

Figure 1.1. Overview of good (a-i) and bad biofilms (j-o): Fixed bed biofilm reactors (a-c), rotating biological contactors (d), and biofilm reactors where the biofilm is suspended in the system (e-g). Biofilms contribute to natural attenuation or controlled bioremediation of ground water and soils (h). In surface waters, benthic biofilms degrade contaminants in the water (i). Unwanted biofilms are a problem in biological fouling of heat exchanges (j) and membrane systems (k). In the medical field, biofilms are related to dental hygiene (l), to infectious diseases (i.e., cystic fibrosis), and problems related to medical implants. In drinking water, biofilms can lead to water quality degradation (m). Increased drag forces can be the result of biofilms growing on ship hulls (n) and growth of biofilms on metal surfaces can lead to biologically induced corrosion (o).

Good biofilms also are ubiquitous in the environment. For example, bacteria in the subsurface normally grow as biofilms on the soil matrix (Figure 1.1h) and can help remove contaminants from the soil or ground water. Pumping nutrients, electron donors, or electron acceptors into the soil can enhance *in situ* biodegradation of contaminants. Natural attenuation of contaminants can occur if environmental conditions in the soil are favorable for biofilm development and metabolic reactions leading to contaminant destruction or immobilization (National Research Council 2000). In rivers, lakes, and coastal areas of the sea, a large fraction of bacterial activity is located in biofilms colonizing stones and sediments (Figure 1.1i). Biofilms also occur naturally in soils and on the roots of plants. All these naturally occurring biofilms are crucial for cycling nutrients in the Earth's biosphere.

Bad biofilms occur in many situations. For instance, biofilms are major problems in dental hygiene (Figure 1.1l), infectious diseases (e.g., cystic fibrosis), and infections related to medical implants (e.g., catheters, heart valves, contact lenses). Growth of biofilms in drinking-water distribution systems is another example of unwanted biofilms (Figure 1.1m). Using organic matter and ammonia present in treated water, bacteria form biofilms in distribution-system pipes. The biofilms and their metabolic reactions cause the water quality to deteriorate in terms of public health and aesthetics. Reduced heat or mass transfer can be the result of biofilms growing on heat exchangers, condenser, and membranes (Figure 1.1j,k).

Bad biofilms often cannot be prevented, as they develop even under adverse conditions (extreme pH values, temperatures up to 95°C, high shear conditions, or disinfectants). Removing biofilms is often difficult in technological systems without a direct access to the exposed surfaces. One prime example is a membrane system that is used for water purification and is prefabricated in spiral-wound modules. Once biofilms develop in these modules, they generally cannot be cleaned again and often need to be discarded. Unwanted biofilms growing on ship hulls (Figure 1.1n) increase the drag forces resulting in a significantly decreased fuel efficiency of these ships. Finally, biofilm growth on metal surfaces has been shown to be a major factor in promoting corrosion (Figure 1.1o).

Overall, biofilms play significant roles in many natural and engineered systems. Understanding the mechanisms of biofilm formation, growth, and removal is the key for promoting good biofilms and reducing bad biofilms. The two definitions at the beginning of this section underscore that a biofilm can be viewed simply or by taking into account complexities. The "better" definition depends on what we want to know about the biofilm and what it is doing. Mathematical modeling is one of the essential tools for gaining and applying this kind of mechanistic understanding of what the biofilm is and is doing.

1.3 WHAT IS A MODEL?

A mathematical model is a systematic attempt to translate the conceptual understanding of a real-world system into mathematical terms (National Research Council 1990). A mathematical model is only as good as the conceptual understanding of the processes occurring in the system. If the conceptual model is good, the model should reproduce the relevant phenomena. If it is not good, we know that we must improve the conceptual model. Thus, a model is a valuable tool for testing our understanding of how a system works.

Creating and using a mathematical model require six steps.

1. The important variables and processes acting in the system are identified. As a simple example for biofilms, substrate and active biomass are variables, and processes can include utilization and diffusion for substrate and synthesis and decay for active biomass.

2. The processes are represented by mathematical expressions. Continuing the biofilm example, substrate utilization can be represented by Monod kinetics and diffusion by Fick's law.

3. The mathematical expressions are combined together appropriately in equations that express balances on mass, energy, or momentum. Again, for the same simple example, the mass balances are on substrate and active biomass at any position inside the biofilm.

4. The parameters involved in the mathematical expressions are given values appropriate for the system being modeled. For example, substrate utilization involves the maximum specific utilization rate and the Monod half-maximum-rate concentration.

5. The equations are solved by a technique that fits the complexity of the equations. Very simple systems can be solved with purely analytical solutions, but numerical solution techniques are needed for systems that are more complicated.

6. The model solution outputs properties of the system that are represented by the model's variables. For example, the model may output the concentrations of substrate at all positions in the biofilm and the flux of substrate into the biofilm's outer surface.

Modeling is a powerful tool for studying biofilm processes, as well as for understanding how to encourage good biofilms or discourage bad biofilms. The main reason is that biofilms naturally have complex interactions of microbiological, physical, and chemical processes. Even the simplest, most homogeneous biofilm develops concentration gradients from the interplay of diffusion with utilization. When a biofilm has complex physical and microbiological structures, many more processes interact. A mathematical model is the perfect means to connect the different processes to each other and to weigh their relative contributions.

1.4   THE RESEARCH CONTEXT FOR BIOFILM MODELING

Mechanistically based modeling of biofilms began in the 1970s (e.g., Williamson and McCarty 1976; Harremoës 1976; Rittmann and McCarty 1980). The early efforts focused mainly on substrate flux from the bulk liquid into the biofilm. The mathematical model represented the biofilm as a simple "slab" in which substrate gradients are in one dimension, perpendicular to the substratum (i.e., the surface onto which the biofilm is attached). Experimental measurements were of the overall substrate-removal rate and the total biofilm accumulation.

Today, experimental techniques available for a detailed evaluation of structure and activity of biofilms have advanced significantly. Microsensors can be used to measure concentrations of many soluble compounds directly within the biofilm (e.g., oxygen, ammonia, nitrate, sulfide, pH). Thus, the availability of substrates and electron acceptors in different regions of the biofilm can be evaluated (Zhang and Bishop 1994a). Rapid advances in molecular biology and *in situ* hybridization techniques have resulted in the development of gene probe and microscopy techniques that permit the detailed analysis of microbial communities in complex biofilms (Lawrence *et al.* 1994; De Beer *et al.* 1997; Silyn-Roberts and Lewis 1997). For example, strain-specific and group-specific ribosomal RNA (rRNA)-targeted probes and confocal laser scanning microscopy (CLSM) are used to investigate the ecology of several diverse types of biofilms, the rumen of animals, activated sludge, and sulfate-reducing fixed-bed reactors (Stahl *et al.* 1988; Amann *et al.* 1992). Similarly, fluorescently labeled antibodies are used to examine natural microbial communities in complex environments such as soils or natural waters (Bohlool and Schmidt 1980). Techniques to study the spatial structure of biofilms are taking advantage of histological tools, such as micro-slicers (Zhang and Bishop 1994b).

With the application of these new techniques and tools, new experimental models to grow biofilms in the laboratory have also been developed. Examples are flow cells that can be directly placed on the stage of a microscope and used to observe biofilm development in real time. However, the use of flow cells is in most cases restricted to the initial stages of biofilm development (experiments generally shorter than 2 weeks and usually less than a few days) and to thin biofilms (conventional confocal laser scanning microscopy does not allow to image biofilms thicker than 100 μm). Flow cells are good examples of laboratory model systems to study certain features of biofilms in great detail but they neglect other features and operating conditions.

Motivated by the new experimental discoveries and enabled by increasingly powerful computers and numerical methods, mathematical models have evolved in parallel (Noguera *et al.* 1999a). The visualization of heterogeneous structures in biofilms (e.g., using images from confocal laser scanning microscopy) has triggered the development of a new generation of mathematical models in which the three-dimensional structure of the biofilm is simulated. The ability to perform *in situ* visualization of individual micro-colonies within a biofilm has fueled the creation of biofilm models that reproduce multi-species interactions. Because of the flexibility offered by modeling and because of the potential to integrate a multitude of processes into a single computational unit, mathematical modeling is becoming a more important tool in biofilm research.

1.5   A BRIEF OVERVIEW OF BIOFILM MODELS

Mathematical models come in many forms that can range from very simple empirical correlations to sophisticated and computationally intensive algorithms that describe three-dimensional biofilm morphology and activity. The best choice depends on the type of biofilm system studied, the objectives of the model user, and the modeling capability of the user.

An example of such a very simple empirical approach is shown in equation (1.1), which describes the BOD$_5$-removal efficiency of trickling filters in wastewater treatment:

$$\textit{Efficiency in } \% = \frac{100}{1 + 0.505 \left(\dfrac{B_{V,BOD}}{F} \right)^{0.5}} \tag{1.1}$$

where $B_{V,BOD}$ is the BOD$_5$ load per filter volume in kg/m^{3}d, and F is the ratio of the flow rate approaching the trickling filter and the wastewater flow (National Research Council 1946). Like most empirical models, this one is based on finding patterns from a large quantity of data obtained under relevant operating conditions. An empirical model can be used to estimate the performance of similar systems as long as the operating conditions are within the range of the evaluated data. Most empirical models provide little insight into biofilm mechanisms, and they should not be used to predict performance outside the tested range of conditions.

Starting in the 1970s, several mathematical models were developed to link substrate flux into the biofilm to the fundamental mechanisms of substrate utilization and mass transport (Harris and Hansford 1976; Harremoës 1976; LaMotta 1976; Williamson and McCarty 1976; Rittmann and McCarty 1980; Rittmann and McCarty 1981). The major goal of these first-generation mechanistic models was to describe mass flux into the biofilm and concentration profiles within the biofilm of one rate-limiting substrate. The models assumed the simplest possible geometry (a homogeneous "slab") and biomass distribution (uniform), but they captured the important phenomenon that the substrate concentration can decline significantly inside the biofilm.

Starting in the 1980s, mathematical models began to include different types of microorganisms and non-uniform distribution of the biomass types inside the biofilm (Kissel *et al.* 1984; Wanner and Gujer 1984, 1986; Rittmann and Manem 1992). These second-generation models still maintained a simplified one-dimensional geometry, but spatial patterns for several substrates and different types of biomass were added. A main motivation for these models was to evaluate the overall flux of substrates and metabolic products through the biofilm surface.

Beginning in the 1990s, and carrying onto today, new mathematical models are being developed to provide mechanistic representations for the factors controlling the formation of complex two- and three-dimensional biofilm morphologies (Wimpenny and Colasanti 1997; Hermanowicz 1998; Picioreanu et al. 1998a, 1998b, 2001, 2004; Noguera et al. 1999b; Kreft et al. 1998, 2001; Eberl et al. 2001; Pizarro et al. 2001; Laspidou and Rittmann 2004; Xavier et al. 2005a). Features included in these third-generation mathematical models usually are motivated by observations made with the powerful new tools for observing biofilms in experimental systems (Section 1.4).

Today, all of the model types are available to someone interested in incorporating mathematical modeling into a program of biofilm research or application. Which model type to choose is an important decision. The third-generation models can produce highly detailed and complex descriptions of biofilm geometry and ecology; however, they are computationally intense and demand a high level of modeling expertise. The first-generation models, on the other hand, can be implemented quickly and easily – often with a simple spreadsheet – but cannot capture all the details. The "best" choice depends on the intersection of the user's modeling capability, biofilm system, and modeling goal.

1.6   GOALS FOR BIOFILM MODELING

A researcher or practitioner who chooses to use biofilm modeling may have one or more of the following goals. The chosen model should match the goals as much as possible.

Understand fundamental mechanisms. Modeling is a powerful tool for a researcher to test his/her understanding of the mechanisms fundamental to how a biofilm forms or performs. The explicitly quantitative nature of a model gives structure to a conceptual understanding and it also allows a rigorous evaluation of the understanding against experimental results.

Link different types of mechanisms. A mathematical model is the ultimate tool for integrating different mechanisms occurring a different spatial and temporal scale: e.g., transport, metabolic, chemical, mechanical, and genetic. The quantitative framework makes it possible to combine and compare seemingly disparate phenomena on an equal footing.

Pre-model experimental design. One of the most effective techniques for ensuring that experiments yield the best information is to pre-model the system to generate expected results. Pre-modeling is most effective for complex systems that involve many interacting phenomena. Pre-modeling usually eliminates experimental designs that fail to provide results that test the underlying hypothesis. Furthermore, pre-modeling helps the researcher identify when his/her understanding can be improved, because the experimental results do not follow the pre-modeling expectations.

Create novel process designs. Modeling can be used to evaluate novel process designs without the cost, time, and risk of building a physical prototype of the process. Modeling gives researchers and practitioners the freedom to identify the most promising designs for experimental testing, while discarding those that cannot meet performance goals.

Improve the performance of a process. The effectiveness of a wide range of operating strategies can be tested for a process, whether it is already build or still in the concept stage. This approach efficiently screens many options without the cost, time, and risk of implementing all the strategies. The most promising strategies can be selected for testing, while ineffective strategies are discarded.

Depending on the characteristics of the biofilm system considered and the questions to be evaluated with the model, different levels of model complexity are required. Therefore, the selection of an appropriate mathematical model must consider the objective of modeling, the data available, the simplifications imposed, and the consequences of such simplifications. A good general rule in modeling is that a model should be as simple as possible and only as complex as needed.

1.7   THE IWA TASK GROUP ON BIOFILM MODELING

The evolution of experimental and mathematical-model tools provides biofilm researchers and practitioners with constantly improving understanding of biofilm complexity. At the same time, researchers and practitioners wonder where to begin if they wish to apply these powerful tools for biofilm design, operation, or research (Noguera *et al.* 1999a; Morgenroth *et al.* 2000). This challenge was discussed in an IWA-sponsored workshop on biofilm modeling in 1998 (Noguera *et al.* 1999a). Following up on the conclusions and recommendations of that workshop, the IWA formed the Task Group on Biofilm Modeling with the complementary purposes of (1) performing a comparative analysis of different modeling approaches and (2) providing researchers and practitioners with guidance for selecting appropriate models to address their needs. The members of the Task Group are listed and other key contributors are acknowledged at the beginning of the report. This report is the final product from the Task Group on Biofilm Modeling.

1.8   OVERVIEW OF THIS REPORT

1.8.1 Guidance for model selection

The ultimate goal of Chapter 2 is to provide guidance on model selection. To do this, Chapter 2 begins by identifying the features of biofilms that are most relevant when creating a mathematical model. These features must be translated into constituents that can be represented mathematically, and Chapter 2 describes the systematic way that the parts of a model are built up so that they represent the desired feature:
• It is necessary to divide the space that includes the biofilm into distinct compartments, such as the substratum onto which the biofilm accumulates, the biofilm itself, and the overlying bulk water.
• Each compartment contains characteristic dissolved components, such as substrates, and particulate components, such as bacteria.
• The components are consumed or produced through transformation processes, and they move about the biofilm space through transport and transfer processes.
• All the processes must be quantified through rate terms that have quantifiable parameters.
• The rate terms for all the processes that affect a component in a particular compartment must be summed up through a mass balance.
• The mass balance equations must be solved by a numerical or analytical technique, yielding the model output.

This modeling framework makes it possible to distinguish among different models in ways that are relevant for selecting a model. Chapter 2 builds upon this framework by identifying distinctive outputs that discriminate among the modeling options available today. The distinction is mainly between outputs that constitute the fine details of the biofilm versus outputs that are averaged over larger distances. The former are called *microscale outputs,* while the latter are called *macroscale outputs.*

1.8.2 Biofilm models considered by the Task Group

All models included in the work of the Task Group follow the framework outlined above, but they differ in the number of simplifying assumptions on which they are based. The types of biofilm models distinguished in the report are:

Multidimensional numerical models: The biofilm is modeled as two- or three-dimensional structure. Thus, all components can vary in multi-dimensional space, as well as time. These models can generate complex physical and ecological structures. Numerical solutions require more computing power than for one-dimensional models, but in general, they are now feasible on common personal computers. A distinction is made between the treatment of dissolved and particulate components. The solute fields can be solved by finite-difference (Picioreanu *et al.* 1998a, 1998b, 2001, 2004; Noguera *et al.* 1999b; Eberl *et al.* 2001; Laspidou and Rittmann 2004) or discrete methods (Pizarro *et al.* 2001; Noguera *et al.* 2004). The dynamics of spatial distribution of particulates is computed using cellular automata (Picioreanu *et al.* 1998a, 1998b; Noguera *et al.* 1999, 2004; Pizarro *et al.* 2001; Laspidou and Rittmann 2004), individual based (Kreft *et al.* 2001; Picioreanu *et al.* 2004), or continuum methods (Eberl *et al.* 2001). Software for solving multidimensional models is currently mainly custom made, but simulation programs are beginning to become public (Xavier *et al.* 2005a).

One-dimensional numerical models: All quantities are averaged in the plane parallel to the substratum (i.e., one dimensional), but gradients perpendicular to the substratum can be calculated for all components (Kissel *et al.* 1984; Wanner and Gujer 1986; Wanner and Reichert 1996). The model equations have to be solved numerically, but simulation software is readily available (e.g., Reichert 1998a, 1998b).

Pseudo-analytical models: The pseudo-analytical solutions (e.g., Sáez and Rittmann 1992) are based on one-dimensional numerical solutions solved on a personal computer. However, the outputs of the numerical solutions are transformed to a set of algebraic equations so that the user of the pseudo-analytical solution avoids numerical solution. The pseudo-analytical solutions were developed originally for one microbial species and one rate-limiting substrate.

Analytical models: Analytical models employ simplifying assumptions so that the flux of dissolved substrates into the biofilm can be calculated easily by hand or with a spreadsheet (Pérez *et al.*, 2005); thus, computationally intensive numerical treatment is not needed. They are inherently one-dimensional and for a single species and substrate.

Chapter 3 describes the details of each of the model types. Table 1.1 shows the codes by which each type is identified and the sections in which they are described in Chapter 3.

Table 1.1. Major biofilm model classes considered by the Task Group

Type of model	Code of model	Described in section
Analytical steady state	A	3.2
Pseudo-analytical	PA	3.3
1-dimensional numerical dynamic	N1	3.4
2-dimensional numerical dynamic	N2	3.6
3-dimensional numerical dynamic	N3	3.6

1.8.3 Benchmark problems

One objective of the Task Group was to compare the performance of various biofilm models available. Multi-dimensional dynamic models describe more aspects of the behavior of biofilm systems than do simple models. However, the multi-dimensional models also require more input data and lead to higher computational costs. At issue is whether the increased effort is justified. In order to investigate this issue, the Task Group created three benchmark problems to be solved by various types of biofilm models. The benchmarks represent typical situations in reactor systems used to treat water. However, since the benchmark problems were designed to evaluate the models according to their ability to represent fundamental features of a biofilm system, the trends in the benchmarks are also applicable to biofilms in nature and in situations in which biofilm is unwanted.

Benchmark problem 1 (BM1): A single-species biofilm with a fixed amount of biomass is present on a flat surface in a completely mixed reactor. The main outputs are the average flux of organic substrate (electron donor) and oxygen (electron acceptor) and the concentrations of these substrates in the bulk liquid. BM1 tests the models for their ability to represent a biofilm in which physical and ecological complexities are not relevant.

Benchmark Problem 2 (BM2): A single-species biofilm is present in a channel with a defined flow-velocity profile and biofilm surface texture. The goal is to compare the ability of the models to output substrate flux when the flow field and surface texture are complex. Thus, BM2 tests the models for their ability to represent physical complexity, but not ecological complexity.

Benchmark Problem 3 (BM3): A multi-species biofilm contains heterotrophic, autotrophic, and inactive biomass. The heterotrophs and autotrophs consume different donor substrates, but both decay to form inert biomass. The goal is to compare model outputs for the fluxes of the two substrates and the accumulation and distribution of the three distinct types of biomass. Thus, BM3 tests the models for their ability to represent ecological complexity, but not physical complexity.

Chapter 4 describes each benchmark problem in detail. The results are used to identify which models are effective for representing the different types of complexity that can be found in biofilms.

2

Model selection

2.1 BIOFILM FEATURES RELEVANT TO MODELING

The first step in creating or choosing a biofilm model is to identify the essential features of the biofilm system. Features are organized into a logical hierarchy:

- *Compartments* define the different sections of the biofilm system. For example, the biofilm itself is distinguished from the overlying water.
- Within each compartment are *components*, which can include the different types of biomass, substrates, products, and any other material that is important to the model.
- The components can undergo transformation, transport, and transfer *processes*. For example, substrate is consumed, and this leads to the synthesis of new active biomass.
- All processes affecting each component in each compartment are mathematically linked together into a *mass balance equation* that contains rate terms and parameters for each process.

This chapter systematically describes the features of biofilm models and how they can be represented by rate terms in mass balance equations. A principle underlying the chapter is that the model user must understand the features that could be important in a biofilm model. However, not every model needs to have all possible features. Therefore, the end of the chapter provides guidance to help the model user discriminate between the features that are essential and those that can be omitted without compromising the value of the model for its intended purpose.

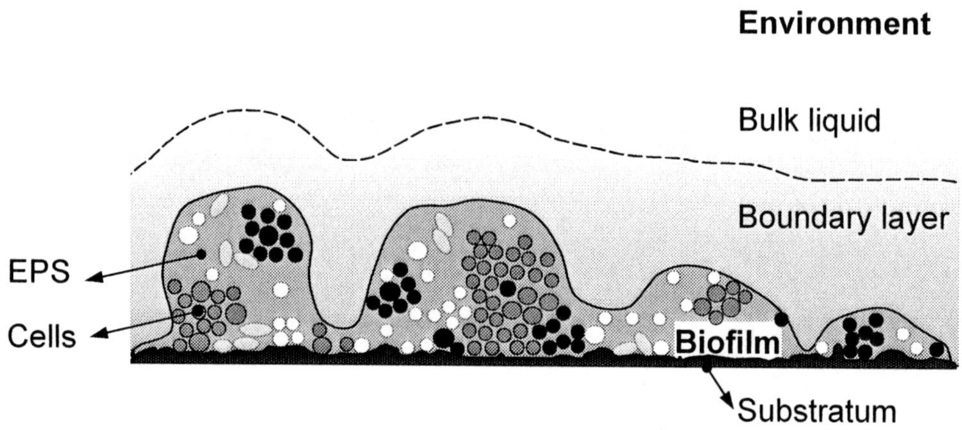

Figure 2.1. Four compartments typically defined in a biofilm system: bulk liquid, boundary layer, biofilm and substratum.

2.2   COMPARTMENTS

A biofilm is a gel-like aggregation of microorganisms and other particles embedded in extracellular polymeric substances (EPS). A biofilm contains water inside it, but its main physical characteristic is that it is a solid phase. A biofilm normally is anchored to a solid surface called the *substratum* on one side and in contact with liquid on its other side. Frequently, a mass-transfer boundary layer is included between the bulk liquid and the biofilm itself. Thus, Figure 2.1 illustrates a biofilm having four compartments: the substratum, the biofilm itself, the boundary layer, and the bulk liquid outside of the biofilm.

2.2.1 The biofilm

The biofilm compartment can be defined in various ways, depending on biofilm structure, system geometry, modeling concept, and modeler's particular needs and goals. While the definition of the biofilm compartment is simple for a flat biofilm surface (Figure 2.2A), it is complex for a heterogeneous biofilm morphology (Figure 2.2B and Figure 2.2C). For example, some consider all material below the maximum biofilm thickness as part of the biofilm compartment (Figure 2.2B), but others consider only the space occupied by the particulate components (i.e., bacterial cells, EPS) as defining the biofilm compartment (Figure 2.2C). Defining the biofilm compartment in the simpler manner of Figure 2.2B is computationally less demanding than using the spatially more complex definition of Figure 2.2C, but the simpler definition dilutes the density of the solids significantly with the water above the solids.

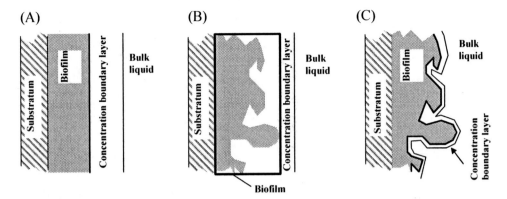

Figure 2.2. Examples of biofilm compartments modeled as a homogeneous structure with simple (planar) geometry (A), a heterogeneous structure with a geometrically simple (planar) biofilm compartment, but with irregularly shaped water pores and channels considered as part of the biofilm compartment (B), or a homogeneous structure where the irregular geometry of the biofilm compartment is strictly defined as the space occupied by the particulate components (C).

The biofilm compartment contains liquid and solids of different types. Although the liquid usually constitutes the majority of the biofilm's mass, the solids are the focus of modeling, since they give the biofilm its reactive and structural properties. The solids include active cells, organic and inorganic particles, and EPS. The liquid in a biofilm contains water found inside the cells, closely associated with the outside of the solids, or more or less free with the spaces between the solids. A model may describe each of the solid and water components separately, or it may treat everything inside the biofilm as only one solid-like component broadly defined as "biomass".

The spatial resolution applied to modeling the biofilm compartment can range from 0-dimensional (0d) to 3-dimensional (3d). With 0d resolution, all biomass components are evenly distributed throughout the compartment; concentration gradients of dissolved components are not present. A 1d model allows property gradients in one direction, usually over the biofilm depth, i.e., from the bulk liquid to the substratum (Figure 2.3A). Often, the spatial resolutions of the solid and liquid phases within the biofilm compartment are modeled differently. For instance, the particulate components in the solid phase can be represented using 0d resolution, while the dissolved components are described with 1d resolution. This hybrid approach recognizes that the concentration gradients of soluble species usually are much stronger than for the solid components.

The 2d and 3d models are needed to reproduce complex spatial distributions of the components (particularly the solid components, as in Figure 2.4A and Figure 2.4B), complex biofilm geometries (e.g., irregular biofilm boundaries at the bulk liquid side such as in Figure 2.2C and Figure 2.5), or complex, small-scale (i.e., μm) geometries at the substratum interface (Figure 2.4C). In these cases, 2d or 3d distributions of the dissolved components are calculated, such as those shown in Figure 2.3B.

When the geometry of the substratum is simple — planar (e.g., biofilms in flow-cells), cylindrical (e.g., biofilms in tubes and pipes), or spherical (e.g., biofilms on support particles in fluid bed reactors) — substratum boundaries can be modeled by 1d approaches. Furthermore, even if the substratum is irregularly shaped at a macroscale level (e.g., crushed stones or plastic media), the length of biofilm compartments used in modeling are usually less than a few millimeters, and therefore, it is acceptable to assume a planar substratum.

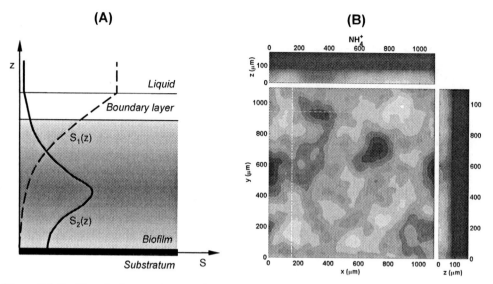

Figure 2.3. Profiles of substrate concentration in biofilms: (A) One-dimensional substrate profiles over the biofilm depth; (B) orthogonal sections of a three-dimensional field of substrate concentration. (Picioreanu *et al.* 2004)

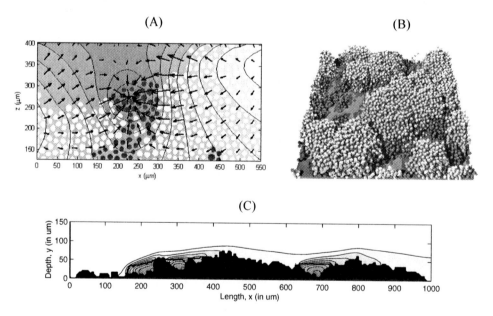

Figure 2.4. Examples of spatial biomass distribution obtained by 2d (A) and 3d (B) simulations of nitrifying multi-species biofilms. White cells are ammonium-oxidizing bacteria (AOB) and black cells are nitrite-oxidizing bacteria (NOB) (Picioreanu *et al.* 2004). Arrows indicate 2d fluxes of intermediate nitrite transported from AOB to NOB. (C) Example of two-dimensional simulation of flow, mass transport and reaction in a biofilm developing on irregular support (shown here as an irregularly shaped black area). (Picioreanu *et al.* 2001)

Day 0   **Day 18**   **Day 25**

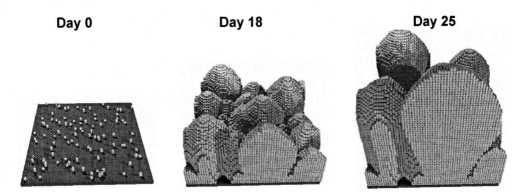

Figure 2.5. Development in time of a biofilm simulated with a 3-dimensional continuum model (Eberl *et al.* 2001).

Time is another dimension that can be considered in biofilm modeling. Whether time needs to be included in a model depends on the characteristic times of the processes taking place in the compartment. Characteristic times often differ by many orders of magnitude. Effects of time on process whose characteristic time is orders of magnitude different from the characteristic times of the most important processes usually can be ignored. For example, the characteristic times of reaction and diffusion processes of soluble components are very short (< 1 second), and time is often ignored for them by solving directly for their steady-state concentrations. At the other extreme, characteristic times for biomass growth and loss processes are large (hours to weeks) and can be ignored when the dynamic growth or loss of biofilm is not at the heart of the modeling goal. When simulating the dynamics of the biofilm biomass is a goal of the modeling, time is a critical dimension that introduces a new challenge: the shape and size of the biofilm compartment change over time (Figure 2.5).

Adding more dimensions (in space or time) to represent greater complexity increases the difficulties in programming the model and in the computational effort of its simulations. Therefore, the wise modeler makes appropriate simplifying assumptions to keep the number of dimensions as small as possible to capture the most important components.

2.2.2 The bulk liquid

In most cases, a bulk-liquid compartment lies over the biofilm compartment. This compartment can be very large as compared to the biofilm, such as for a biofilm growing on the sediment of a lake, or it can be a thin layer of water, as in a trickling filter, soil, or dental plaque. Dissolved and particulate components can exchange between the biofilm and the bulk liquid. For example, the bulk liquid usually supplies nutrients utilized by the microorganisms in the biofilm, while pieces of biofilm solids detach from the biofilm compartment and move to the bulk-liquid compartment.

From the point of view of the biofilm compartment alone, the simplest way is to handle the bulk liquid is as a boundary condition of the biofilm compartment. One very simple way to do this is to specify the concentrations of dissolved components at the interface between the two compartments. This simple approach works well if the biofilm does not affect the concentrations in the bulk liquid. However, the exchange of material between the biofilm and the bulk liquid often has a profound impact on the concentrations in the bulk liquid. Biofilm treatment reactors used to remove pollutants from water are perfect examples of profound impacts. When the effect of the exchange on the bulk liquid is important, the most direct approach is to include the bulk liquid is a separate, completely mixed compartment. In that compartment, component concentrations vary according to the inflow, outflow, conversion reactions in the bulk liquid, and, most significantly here, exchanges with the biofilm. The bulk-liquid compartment still constitutes a boundary condition of the biofilm compartment, but with concentrations that change over time. Likewise, the biofilm is a boundary condition of the bulk liquid.

Detachment of solid components from the biofilm can be considered in all these models, but only in the case when the bulk liquid is simulated as a separate compartment is it possible to calculate the concentration of solids suspended in the bulk liquid. Attachment of solid components from the bulk liquid and onto the biofilm surface can be considered in a similar way.

Sometimes it is important to simulate the changes in the concentration of dissolved components along the direction of the flow. This is most easily accomplished by separating the biofilm system into a number of small sub-systems in series. Each biofilm sub-system has its own compartments, and each bulk-liquid compartment can be completely mixed. For a series of sub-systems in series along the flow path, the components in the bulk liquid transport from upstream to downstream sub-systems with the water. By combining sub-systems in parallel, as well as it series, it is possible to simulate quite complex flow patterns (Fogler, 1999).

A computationally very complex and intense alternative is to simulate the entire domain as a single biofilm system, but use the Navier-Stokes and continuity equations for the simulation of fluid flow and mass balances for the dissolved and suspended components in the bulk liquid compartment. This alternative requires that the spatial resolution of the biofilm and the bulk liquid be 2d or 3d. Since the concentration gradients in the bulk liquid are a model output, an *a priori* definition of the boundary condition between the biofilm and bulk liquid compartments can be avoided.

The question of whether the bulk liquid is described by a steady state or a dynamic model depends on the objective of the modeling and this is closely related with the way the biofilm is treated mathematically.

2.2.3 The mass-transfer boundary layer

Experimental observations clearly indicate strong concentration gradients for solutes just outside the biofilm when these solutes are utilized or produced by the microorganisms in the biofilm. Consequently, the solute concentrations at the biofilm surface and in a completely mixed bulk liquid often are significantly different. Hence, the simplest boundary condition - the solute concentration at the biofilm's outer surface equals its concentration in the bulk liquid - often is incorrect. A more sophisticated boundary condition is needed, and it is achieved by introducing a third compartment, the mass-transport boundary layer.

The mass-transport boundary layer is a hypothetical layer of liquid above the biofilm and in which all the resistance to mass transport of dissolved components outside the biofilm occurs. For biofilms having relatively flat outer surfaces, the boundary layer is typically assumed to be of uniform thickness. On the other hand, biofilms with highly irregular surfaces do not necessarily have a mass-transport boundary layer with a uniform thickness. That case may require that concentrations within the mass-transport boundary layer compartment be computed using 2d or 3d numerical approaches similar to those used for the biofilm compartment.

For 2d and 3d models that model the flow field (via the Navier-Stokes equation) and the concentration gradients in the bulk liquid, the boundary layer concept is not needed, since mass-transport effects near the biofilm surface are included automatically.

In situations with very strong turbulence or high flow velocities in the bulk liquid, concentration gradients outside the biofilm can become insignificant, and the mass-transfer boundary layer may be neglected.

2.2.4 The substratum

The solid surface on which the biofilm grows is called the substratum, and it normally represents a separate compartment in the biofilm system. However, it usually is a very simple compartment to model, because the typical substratum is inert and impermeable. Rocks, sand, and plastic biofilm carriers are good examples of inert substrata. Nothing enters or leaves an inert substratum, and it has no transformations.

Some substrata are not inert. These include organic solids that are biodegraded by attached microorganisms, metal surfaces that dissolve by biocorrosion, dental enamel that can dissolve because of dental plaque, and plants that excrete substrates that support biofilm growth. In other situations, the substratum may be a permeable or semi-permeable membrane that supplies the biofilm with gaseous or dissolved substrates. In all these cases, the substratum compartment should be represented as a reactive boundary, much like the bulk liquid is an active boundary.

2.2.5 The gas phase

In certain cases dissolved components in the bulk liquid exchange with an adjacent gas phase, O_2 for aerobic respiration being the most common example (other gases include CH_4, N_2, H_2, trichloroethene (TCE), and toluene). Gaseous components can be transferred into the liquid, aeration to add O_2 is the most common example. Volatilization of low-solubility components, like TCE and CH_4, decreases their bulk liquid concentrations. In all cases, the exchanges between the gas and liquid phases require boundary conditions. When the exchange affects the concentration of the gas in the gas phase, the gas phase must be included as a separate compartment of the biofilm system.

2.3 COMPONENTS

Components in biofilm modeling are divided into particulate and dissolved. Particulate components are the materials that form the biofilm solid phase, such as cells and EPS. They are physically attached to each other or to the substratum. However, particulate components such as cells can be found also suspended in the bulk liquid. Dissolved components are the dissolved species found in the biofilm liquid phase, such as substrates and metabolites.

The distinction between particulate and dissolved components is important because the characteristic times of the processes related to the two types of components are significantly different. Generally, processes associated with the particulate components are much slower than for the dissolved components, and their characteristic times are correspondingly larger. Model simulations can be made computationally more efficient by separating the processes associated with the dissolved and particulate components. A second way that the distinction is useful is that the particulate components do not diffuse, while soluble components do.

2.3.1 Dissolved components

Dissolved components include substrates, metabolic intermediates, and various products of microbial conversion processes. A common assumption is that, for one microbial metabolic activity within the biofilm, of the rate of growth is controlled by the concentration of one rate-limiting substrate. Utilization of all substrates and nutrients, and the generation of products, is then proportional to the growth rate. As long as the concentrations of other dissolved components do not become rate limiting at any location within the biofilm, they do not need to be considered as rate-limiting variables. However, when it cannot be guaranteed that only one substrate is rate limiting throughout the entire biofilm compartment, the concentrations of multiple substrates must be solved simultaneously, a step that increases modeling complexity and computing demands, since a mass balance must be solved for each rate-limiting variable. Example 1 illustrates how the decision to use one or multiple substrates can be handled in a biofilm compartment.

Multiple substrates controlling the metabolic activity of a single type of microorganism must be distinguished from multiple substrates needed to support more than one microbial metabolic type. For example, aerobic heterotrophs oxidize an organic substrate and reduce oxygen, while nitrifiers oxidize ammonium and reduce oxygen. When heterotrophs and nitrifiers co-exist in a biofilm (a common situation), growth and substrate utilization of the heterotrophs can be controlled by the concentrations of the organic substrate, oxygen, or both. Likewise, the growth and substrate utilization of the nitrifiers can be controlled by the concentration of ammonium, oxygen, or both. If oxygen is not limiting for either microbial type, the model still must include two substrates, one for each microbial type: organic substrate for the heterotrophs and ammonium for the nitrifiers. However, multiple substrates are included (as in Example 1) for a microbial type only when oxygen also becomes rate limiting for that microbial type.

Example 1: *When do multi-substrate limitations have to be considered for a single microbial metabolic type in a biofilm?*

Question:
Biological conversion processes require two substrates, an electron donor and an electron acceptor. Although biofilm models may need to consider transport and conversion of both substrates, in some cases one compound fully penetrates the biofilm so that mass transport needs to be considered only for the other substrate. Solving the diffusion-reaction equations is greatly simplified if only one limiting substrate has to be considered. How can a modeler evaluate *a priori* whether or not both substrates need to be considered?

Answer:
The utilizations of electron donor and electron acceptor are linked to bacterial growth. As an example, we can consider the aerobic oxidation of organic substrate. The rate of heterotrophic growth (r_g) and the associated rates of substrate (r_S) and oxygen (r_{O2}) utilizations are

$$r_g = \mu_{max,H} \frac{S_S}{K_S + S_S} \frac{S_{O2}}{K_{O2} + S_{O2}} X_H \qquad (2.1)$$

$$r_S = \frac{1}{Y_H} r_g \qquad (2.2)$$

$$r_{O2} = \frac{(1 - Y_H)}{Y_H} r_g \qquad (2.3)$$

[Note that Eq. (2.3) only takes into account growth-associated oxygen utilization and neglects endogenous respiration. Oxygen also can be consumed for endogenous respiration, a factor considered later in the book. The simplification used in Eq. (2.3) is appropriate for situations in which r_g is significantly greater than zero.]

Dividing Eq. (2.2) by Eq. (2.3) gives the stoichiometric coefficient linking substrate and oxygen utilizations:

$$v_{S,O2} = \frac{r_S}{r_{O2}} = \frac{1}{(1-Y_H)} \tag{2.4}$$

Substrate penetration also depends on the diffusion coefficients (D_S, D_{O2}) and concentrations at the surface of the biofilm for organic substrate (S_S) and oxygen (S_{O2}), respectively:

$$\gamma_{S,O2} = \frac{1}{v_{S,O2}} \frac{D_S}{D_{O2}} \frac{S_{LF,S}}{S_{LF,O2}} \tag{2.5}$$

where $\gamma_{S,O2}$ is the penetration of organic substrate relative to the penetration of oxygen (Henze *et al.* 2002). Three cases can be differentiated using $\gamma_{S,O2}$:

$\gamma_{S,O2} \gg 1$ Oxygen is potentially limiting inside of the biofilm, but organic substrate is fully penetrating the biofilm.

$\gamma_{S,O2} \approx 1$ Not clear whether oxygen or organic substrate is limiting inside of the biofilm. A multi-substrate model should be used to evaluate the biofilm.

$\gamma_{S,O2} \ll 1$ Organic substrate is potentially limiting inside of the biofilm, but oxygen is fully penetrating the biofilm.

Concentration profiles for organic substrate and oxygen are shown in Figure 2.6 for different values of $\gamma_{S,O2}$ (Table 2.1). Results in Figure 2.6 are based on numerical solutions of a multi-substrate and multi-species biofilm model and processes, and parameters are as defined in BM3 (Section 4.4). For very large or very small values of $\gamma_{S,O2}$, oxygen or organic substrate is limiting throughout the depth of the biofilm, respectively (cases a, c). However for $\gamma_{S,O2} = 1$, organic substrate and oxygen are significantly decreasing inside of the biofilm, and a multi-substrate model should be used (case b). Note that in case (b) the organic substrate penetrates deeper into the biofilm than oxygen even though $\gamma_{S,O2}$ in Table 2.1 equals 1. This is because equation (2.4) is simplified in that it neglects decay processes.

Thus, the answer to the original question is that the non-dimensional parameter γ can be used to evaluate whether electron donor or acceptor can be assumed to fully penetrate the biofilm. For this compound, mass transport does not need to be calculated explicitly, and a single-substrate model can be used when γ is not near 1. If γ is approximately equal to 1, each substrate must be included.

Table 2.1. Bulk phase concentrations for three different cases considered in Figure 2.6.

Case	Concentration at the biofilm surface		Relative penetration
	Oxygen, mg/L	COD, mg/L	$\gamma_{S,O2}$
a	10	5	0.1
b	10	54	1
c	4	54	10

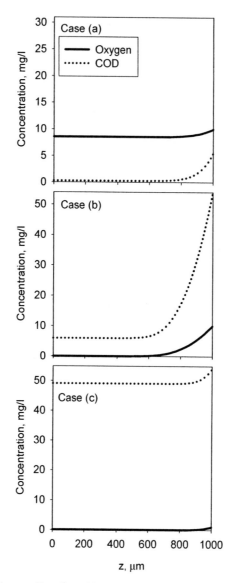

Figure 2.6. Concentration profiles for a 1000-μm thick biofilm for a heterotrophic biofilm with concentrations at the biofilm surface as shown in Table 2.1 and calculated using a multi-substrate numerical solution and the kinetic and stoichiometric parameters as defined in BM3. Cases (a) and (c) show full penetration of the non-limiting compounds for $\gamma_{S,O2}$ much larger or smaller than 1, respectively. Case (b) shows the result for $\gamma_{S,O2} = 1$, and oxygen and organic substrate are fully depleted inside the biofilm.

2.3.2 Particulate components

Particulate components include active microbial cells, dead or inert cells, EPS, and any other organic or inorganic particles embedded in the biofilm solid matrix. Most models described in the literature do not include all of the particulate components. One reason for including only some of the particulate components is that the modeler is not interested in representing the omitted components. For example, only the active cells may be of interest for predicting the substrate flux into the biofilm. A second reason is that the modeler does not have a basis for describing the component. Until recently, models included only microbial cells (active and inactive) as the particulate components, but approaches to simulate the EPS as a separate particulate component have been recently described (Kreft and Wimpenny, 2001; Laspidou and Rittmann, 2004; Xavier et al., 2005a).

In addition to choosing which particulate components to include, the modeler needs to decide about whether the components change with time and how they are distributed in space. The choices depend on the goal of the model and the amount of computing resources that can be applied. When the concentrations of the particulate components or the total amount of biofilm change over time, the model is called dynamic. A dynamic model is essential when the modeler is interested in predicting how the biofilm evolves over time. On the other hand, a static-biofilm model is suitable when the modeler is interested in the substrate gradient or flux for a known amount and type of biofilm. For a static model, the amount of biofilm is not predicted, but is an input to the model. A special case is a steady-state biofilm, in which the amount of biofilm growth is just balanced by its losses; a steady-state-biofilm model is a hybrid between static and dynamic.

When more than one component is present, the modeler must decide how they will be distributed in space. The simplest case is to impose that they are homogeneously mixed in the same proportion in all parts of the biofilm; this approach is the simplest to program and solve. The most complex approach is to have the model predict the distribution by solving the mass balances for all particulate species simultaneously throughout the biofilm. An intermediate approach is to segregate the different types of biomass with a pre-assigned pattern, such as in layers. Going from 3d to 2d to 1d automatically reduces the spatial complexity and reduces the computing demands.

One final consideration about modeling the particulate components is deciding the degree to which the model represents the component as a continuum or as individual cells. The continuum approach, the traditional way to solve the mass-balance equations, does not describe the behavior of individual cells. Instead, the continuum approach provides an average concentration of the component. Phenomena that occur at the scale of single cells can be modeled using individual-based particles or cellular automata.

2.4 PROCESSES AND MASS BALANCES

The two universal types of processes occurring in a biofilm system are *transformations* (e.g. substrate utilization) and *transport* (e.g. diffusion). These two types of processes are intimately related: Substrate utilization inside the biofilm generates gradients that drive diffusive mass transport, and diffusive mass transport supplies the substrates that allow their utilization. Together, substrate utilization and diffusion create substrate gradients. Because microorganisms are exposed to different substrate concentrations, the rates of microbial growth vary with the location within the biofilm. In addition, the gradients can create unique microbial *niches* in different locations in the biofilm. For instance, oxygen may not penetrate all the way to the substratum, allowing stable anaerobic microenvironments in the deep regions of a biofilm.

Another significant class of processes is *transfer* between the biofilm and the bulk liquid. Transfer processes include the exchange of dissolved chemicals across the mass-transfer boundary layer, attachment of planktonic cells from the bulk liquid to the biofilm, and detachment of particulate components from the biofilm to the bulk liquid. In mathematical terms, they are boundary conditions for the mass-balance equations in each compartment.

Several transformation, transport, and transfer processes occur simultaneously for any component. The concentration of the component depends on how all the processes "add up." The tool for adding up the impacts of all processes is the *mass balance*, which is the foundation for any mathematical biofilm model.

For any component, a general mass balance states that, for any time and location in the compartment,

$$\{\text{Net rate of } \textbf{accumulation}\} = \begin{array}{c} \{\text{Sum of all rates of } \textbf{transfer and transport}\} \\ + \\ \{\text{Sum of all rates of } \textbf{transformation}\} \end{array}$$

in which accumulation refers to the total mass of the component in the compartment. The first task of mathematical modeling is to set up mathematical expressions for all terms in brackets. The common processes and their rate expressions are discussed in the following sub-sections.

2.4.1 Transformation processes

Transformation processes usually are biochemical reactions that produce or consume one or more components: e.g., consumption of substrate, production of metabolic end-products, microbial growth and decay, and production of EPS. Transformation processes also can include abiotic chemical reactions, such as adsorption of a solute onto the biomass and precipitation of a mineral solid.

In modeling, the transformation processes are represented mathematically with rate expressions, or equations that tell how fast the component is produced or consumed. The rate is proportional to the concentrations of one or more of the components and one or more kinetic parameters. Depending on how the model is constructed, the rate expressions can have different units, such as mass per time (MT^{-1}), mass per volume per time ($ML^{-3}T^{-1}$), mass per area per time ($ML^{-2}T^{-1}$), mass per mass of biomass per time, ($MM_X^{-1}T^{-1}$), or just per time (T^{-1}). Despite having different units for different rate expressions, two things must be consistent about rate expressions. First, the units for all rate expressions must have time in the denominator, or have T^{-1}. This is true because a rate expressed the change in amount with time. Second, when different rate expressions are combined into a mass balance for a component, all the rate expressions must be converted into exactly the same units, so that they can be added directly.

Because biofilms are formed from active microorganisms, the first type of transformation process to consider is for microbial synthesis. In biofilm models, the rates laws used for describing microbial synthesis are similar to the rate expressions used in the simulation of microbial synthesis of the same microorganisms in suspension. The only difference between a biofilm and a suspended-growth culture in this regard is that the concentrations of the substrates or products affecting the growth rate can vary within the biofilm, and, thus, they are not necessarily the same as the concentrations present in the bulk liquid. For completeness, a brief description of typical rate laws found in biofilm models follows, but the reader can find more extensive descriptions of rate expressions used in the general modeling of biological processes in Rittmann and McCarty (2001) and Grady *et al.* (1999).

A very widely used model describing the synthesis of microbial biomass is the Monod equation for one limiting substrate:

$$\mu = \mu_{max} \frac{S}{K_S + S} \tag{2.6}$$

where μ is the specific growth rate (T^{-1}), μ_{max} the maximum specific growth rate (T^{-1}), S is the concentration of the rate-limiting substrate ($M_S L^{-3}$), and K_S is the concentration ($M_S L^{-3}$) giving one-half the maximum rate. The specific growth rate (μ) is the mass per volume rate ($M_X L^{-3} T^{-1}$) divided by the biomass concentration ($M_X L^{-3}$). It is common in microbiological modeling to divide rates by the biomass concentration; thus, specific rates are common. In the case of μ, the specific growth rate has units of T^{-1}.

The Monod equation is attractive because it captures observed effects of substrate concentration on the rate of biomass synthesis. First, the rate approaches zero when the concentration of the limiting substrate approaches zero. Second, the rate approaches a finite maximum value for large concentrations of S. Finally, the transition between the two extremes is smooth, which is advantageous for numerical treatment.

In some biofilm models, alternative expressions are used, such as the first-order kinetics

$$\mu = k_1 S \tag{2.7}$$

and the zero-order kinetics

$$\mu = k_0 \tag{2.8}$$

which are special cases of Equation (2.6) for $S \ll K_S$ with $k_1 = \mu_{max}/K_S$ and for $S \gg K_S$ with $k_0 = \mu_{max}$.

When multiple substrates are rate limiting, the Monod equation typically is extended to include the effects of each substrate influencing the rate of microbial synthesis. For instance, is the rate of microbial synthesis may be limited by the concentrations of the electron donor (S_1, e.g. COD) and the electron acceptor (S_2, e.g. O_2); then the specific growth rate can be described with a multiplicative-Monod expression (Bae and Rittmann 1996):

$$\mu = \mu_{max} \frac{S_1}{K_{S1} + S_1} \frac{S_2}{K_{S2} + S_2} \tag{2.9}$$

Using the same principle, the effect of an inhibitory component can be modeled by the addition of a specific term describing the effect of the concentration of the inhibitor on the specific growth rate, such as in

$$\mu = \mu_{max} \frac{S}{K_S + S} \frac{K_I}{K_I + I} \tag{2.10}$$

where I is the concentration of the inhibitor ($M_I L^{-3}$) and K_I is the concentration ($M_I L^{-3}$) of the inhibitor giving 50% inhibition of the rate.

Regardless of the expression used for the specific rate, the most common way to express the rate in the mass balance is the volumetric production rate r_X ($M_X L^{-3} T^{-1}$),

$$r_X = \mu X \tag{2.11}$$

where X is the concentration ($M_X L^{-3}$) of the microbial species. As for biomass production, similar rate equations are also used for biomass decay processes (Table 2.2).

Synthesis is not the only transformation process relevant to microorganisms in a biofilm. Indeed, the microorganisms undergo a number of loss processes, including endogenous decay (or inactivation), predation, and detachment. The inactivation process is common in almost any biofilm and serves as a model for how to express a biomass-loss process.

Table 2.2. Simple stoichiometric matrix for heterotrophic synthesis and inactivation

Process	Component		Process rate ρ_j
	X_H	S_S	
1. Heterotrophic synthesis	1	$-\dfrac{1}{Y_H}$	$\mu_{max,H}\dfrac{S_S}{K_S+S_S}X_H$
2. Inactivation of heterotrophic cells	-1		$b_H X_H$

A simple and common means to represent inactivation/decay is with a first-order loss rate:

$$r_{in} = -bX \qquad (2.12)$$

where r_{in} is the inactivation rate ($M_X L^{-3} T^{-1}$) and b is the first-order inactivation rate constant (T^{-1}). The rate has a negative sense, which means that it is causing a loss of active biomass.

Because utilization of a substrate is what brings about biomass synthesis, the rate of substrate utilization (r_S in $M_S L^{-3} T^{-1}$) is proportional to r_X:

$$r_S = -\frac{1}{Y}r_X \qquad (2.13)$$

where Y is the true-yield coefficient ($M_X M_S^{-1}$), or the ratio of biomass produced per unit of substrate consumed. It is important to note that r_X is positive and r_S is negative.

Table 2.2 summarizes the basic rate expressions for heterotrophic bacteria (subscript H) in a convenient and commonly used matrix format. The far-right column gives the form of the rate basic rate expression (i.e., ρ_j) for the j-th process listed by name in the far-left column. The entries in the columns are the stoichiometric multipliers (v_{ij}) of the basic rate expression to give the rate appropriate for the component in the i-th column.

As the model becomes more complex, it is increasingly efficient to represent the transformation processes in the form of a stoichiometric matrix, as in Table 2.2. Rows are added to account for additional transformation processes, and columns are added for more components. The overall (or net) transformation rate for each component r_i ($M_i L^{-3} T^{-1}$), which is inserted into the mass balance equation, can then be calculated directly as

$$r_i = \sum_j v_{ij}\rho_j \qquad (2.14)$$

where v_{ij} denotes the stoichiometric coefficient of component i in process j. For example, component X_H has

$$r_H = \mu_{max,H}\frac{S_S}{K_S+S_S}X_H - b_H X_H \qquad (2.15)$$

and component S_S has

$$r_S = -\frac{1}{Y_H}\mu_{max,H}\frac{S_S}{K_S+S_S}X_H$$

For each component, the units on the rate terms used in the mass balance must be the same, such as mgCOD/L-day for X_H in Table 2.2.

In addition to the microbially catalyzed transformation reactions, a biofilm model may include abiotic chemical transformations. When an abiotic reaction for the component is roughly as fast as for the microbially catalyzed reactions, the abiotic reaction is added to the matrix and handled in a similar manner. In some cases, the abiotic reactions are very much faster than the biological reactions. For instance, acid-base reactions may be included in models to allow the calculation of pH at different locations within the biofilm. Acid-base reactions are infinitely fast and can be modeled as chemical equilibrium reactions (Picioreanu and van Loosdrecht 2002). It also is possible to treat very fast reactions with the formalism used in Table 2.2. A forward reaction rate constant k_f is given an arbitrary value several orders of magnitude higher than the microbial rate constants for the component. This value is then used to calculate the backward reaction rate constant k_b as $k_b = k_f K_{eq}$, where K_{eq} is the equilibrium constant, which is usually known (Wanner 2002). Which of the two approaches to modeling fast transformation processes is to be preferred depends on the way the model equations are treated mathematically.

2.4.2 Transport processes

The transport processes that regularly are considered in biofilm models are advection, molecular diffusion, and turbulent dispersion. In special cases, transport of charged components by migration in an electric field created is included. The general, 1d expression to model the specific mass flux of a component in the direction z (usually used to denote the spatial direction perpendicular to the substratum) is

$$j_z = u_z C - D\frac{\partial C}{\partial z} - D_T\frac{\partial C}{\partial z} - \zeta DC\frac{F}{RT}\frac{\partial \Phi}{\partial z} \qquad (2.16)$$

where j_z is the mass flux of the component ($ML^{-2}T^{-1}$); u_z is the advective velocity (LT^{-1}) in the direction z; D and D_T are the coefficients of molecular diffusion and turbulent dispersion (L^2T^{-1}), respectively; ζ is the ion charge ($\bar{e}N^{-1}$); F is the Faraday constant ($IT\bar{e}^{-1}$); Φ is the electrical potential ($L^2MT^{-3}I^{-1}$); T is the temperature (θ); and R is the universal gas constant ($L^2MT^{-2}\theta^{-1}N^{-1}$). For a soluble ion, all four terms on the right-hand side can act. For a soluble, uncharged component, the fourth term is not relevant. For a particulate component, only the first term (advection) is normally included in the flux definition, since the other terms normally are small. In a multi-dimensional model, analogous expressions for material fluxes in x and y directions, j_x and j_y respectively, have to be specified and substituted into Equation (2.16).

The dominant transport process for dissolved components inside the biofilm compartment usually is molecular diffusion. In most cases, the diffusion coefficient of a component in the biofilm's liquid phase is smaller than in the bulk liquid compartment. This can be attributed to the fact that j_z is defined as the mass flux per unit biofilm area, but the biofilm contains liquid and solid phases. If a component is transported in the biofilm's liquid phase only, the effective area for diffusion is smaller than in the bulk liquid. This is taken into account by a smaller, effective diffusion coefficient, which can be a function of biofilm porosity and pore tortuosity. The effective diffusion coefficient tends to be inversely proportional to the biomass density. Consequently, the diffusion coefficient in the biofilm shows large variability (Beuling et al. 2000). Advection of water inside the biofilm also is possible when large channels are present inside the biofilm compartment.

Outside the biofilm, advection and turbulent dispersion are the dominant transport processes. Advection is easily described as the product of the fluid velocity and the component concentration. The flow velocity either can be an input to the model or can be calculated from the continuity and momentum equations for the fluid (computational fluid dynamics, CFD). On the other hand, turbulent dispersion demands a value for D_T, which is usually estimated from empirical correlations that are system specific. In some situations, mixing in the bulk liquid is so vigorous that spatial gradients of the component concentration can be neglected. Then, transport within the bulk liquid compartment does not have to be explicitly modeled.

Near the biofilm surface, spatial gradients of the component concentration can seldom be neglected. They can also be rigorously calculated by the continuity and momentum equations for the fluid, but usually they are modeled with empirical mass-transfer equations for the biofilm–bulk liquid interface, as is discussed in the next section.

Transport of particulate components that comprise the biofilm occurs through expansion or contraction of the biofilm's solid matrix due to bacterial growth or decay and EPS production. The movement of these components can be represented in various ways. Continuum-based biofilm models consider an advective flux (e.g., Rittmann and Manem 1992, in 1d; Dockery and Klapper 2001, in 2d), a diffusive flux (Eberl *et al.* 2001, in 3d), or both types of fluxes (Wanner and Reichert 1996, in 1d). In models with a discrete description of particulate components, individual cells and particles are displaced according to empirical rules that mimic advective or diffusive flux (Picioreanu *et al.* 1998a, 1998b; Kreft *et al.* 2001; Pizarro *et al.* 2001; Picioreanu *et al.* 2004). In principle, individual-based models can also describe cell-based transport phenomena such as bacterial motility and chemotaxis (Dillon *et al.* 1995).

2.4.3 Transfer processes

Transfer processes exchange mass of dissolved or particulate components between two compartments. At the interface between the compartments, a continuity condition for the component concentration C and the specific flux $\mathbf{j}$ of the exchanged mass must be fulfilled. Continuity means that C and $\mathbf{j}$ are the same on both sides of the interface between the compartments. C and $\mathbf{j}$ can be calculated at each side of the interface from boundary conditions, as is detailed in Section 3.1.1.2 (Eberl *et al.* 2000a; Picioreanu *et al.* 2000a, 2000b).

Mass transfer between the bulk liquid and the biofilm is an especially important process, because the source of substrates in most biofilm systems is the bulk liquid. Substrate utilization in the biofilm requires that it be transported there, and the synthesis and maintenance of the biofilm depends on substrate utilization. Generally, the transport is driven by a concentration gradient across the mass-transfer boundary layer (MTBL) (sometimes called the concentration boundary layer, CBL). The mass flux of a component in the MTBL is proportional to the difference between the component concentration in the bulk liquid, C_B, and at the biofilm surface, C_{LF} (ML^{-3}), with the proportionality constant being the liquid-biofilm mass transfer coefficient k_c (LT^{-1}). The mass flux perpendicular to the biofilm surface j_n ($ML^{-2}T^{-1}$) is

$$j_n = k_c(C_B - C_{LF}) \tag{2.17}$$

The simplest and still most used boundary-layer model, the film theory developed by Nernst, represents mass transfer across the MTBL as molecular diffusion across an effective diffusion layer of thickness L_l. Then, the mass transfer coefficient is $k_c = D/L_l$. For slow flow velocities or low turbulence, the value of L_l is large (high mass-transfer resistance), making k_c small; for fast flow or high turbulence, L_l is small (low resistance), and k_c is large (if the mass-transfer resistance in the boundary layer is extremely low, the boundary layer can be neglected, and $C_B \approx C_{LF}$). The mass transfer coefficient k_c and L_l usually are calculated from experimental correlations (e.g., Perry and Green 1999).

Example 2: *How to estimate the liquid boundary layer or a mass transfer coefficient?*

Question:
Caroline is operating a submerged biofilm reactor and she wants to estimate the extent of external mass transfer limitation. What questions does she need to consider when estimating mass transfer parameters based on equations she finds in the literature? How thick can she expect the external concentration boundary layer in her reactor to be?

System specific information:
   Water flow rate (Q) = 12 m^3/d
   Cross sectional area of the empty reactor = 1 m^2
   Diameter of filter media (d_p) = 4 mm
   Overall bed porosity (φ) = 0.4
   Diffusion coefficient of substrate in water (D) = 10^{-4} m^2/d
   Kinematic viscosity of water (ν) = 1.307·10^{-6} m^2/s

Answer:
When the mathematical model does not explicitly evaluate fluid dynamics, external mass transfer is considered by defining a liquid boundary layer (L_L) or an external mass transfer coefficient ($k_c = D/L_L$). Many empirical correlations are available in the literature for estimating the non-dimensional Sherwood number (Sh):

$$Sh = \frac{k_c d_p}{D} \tag{2.18}$$

where d_p is a characteristic length (e.g., the diameter of a biofilm support particle) and D is the diffusion coefficient in the liquid. The Sherwood number is often expressed as a function of the non-dimensional Reynolds number (Re = $U d_p/\nu$) and Schmidt number (Sc = ν/D_i)

$$Sh = A + B \cdot Re^m Sc^n \tag{2.19}$$

Reynolds number depends on flow and geometry, while Schmidt number only depends on fluid properties. The parameters A, B, m, and n are, in most cases, determined empirically from experimental data for a specific system. For example, for the simple case of fluid flow around a rigid spherical particle $A = 2$, $B = 0.6$, $m = 1/2$, $n = 1/3$ (Frössling, 1938).

Many of the correlations to estimate Sh were derived for relatively simple chemical engineering applications, and they should be applied to biofilm systems with these cautions:

- The parameters A, B, m, and n depend on the geometry of the biofilm support medium and are valid only for a defined range of Re and Sc. Extrapolating from these conditions can yield erroneous results.
- For complex geometries Eq. (2.19) may not be sufficient, and other forms of Sh=f(Re, Sc) may be used to fit experimental data.
- Most correlations for Sh were derived for rigid particles, but the elastic and heterogeneous nature of the biofilm can influence external mass transfer. Different authors have suggested k_c to either decrease (Nicolella *et al.* 2000) or to increase (Horn and Hempel, 2001) because of a biofilm covering the support medium.

Caroline decides to use Eq. (2.19) with the following parameters based on Wilson and Geankoplis (1966): $A = 0$, $B = (1.09 \, \phi^{-2/3})$, $m = 1/3$, and $n = 1/3$, where ϕ is the void fraction of packed bed. These parameters are applicable for $0.0016 < Re < 55$, $950 < Sc < 70,000$, and $0.35 < \phi < 0.75$. For the given system, she calculates the following results:

$$
\begin{aligned}
Re &= (12 \text{ m/d}) \cdot 4 \text{ mm} / (1.307 \cdot 10^{-6} \text{ m}^2/\text{s}) = 0.425 \\
Sc &= (1.307 \cdot 10^{-6} \text{ m}^2/\text{s}) / (10^{-4} \text{ m}^2/\text{d}) = 1{,}129 \\
Sh &= 1.09 \cdot 0.4^{-2/3} \cdot 0.425^{1/3} \cdot 1{,}129^{1/3} = 15.7 \\
k_c &= 15.7 \cdot (10^{-4} \text{ m}^2/\text{d}) / (4 \text{ mm}) = 0.393 \text{ m/d} \\
L_L &= 4 \text{ mm} / 15.7 \quad = 255 \text{ µm}
\end{aligned}
$$

Thus, Caroline can expect L_L to be on the order of 255 µm, or she can use the value of $k_c = 0.393$ m/d. Caroline also verified that ϕ, Re, and Sc are in the applicable ranges for the correlation suggested by Wilson and Geankoplis (1966). Clearly, external mass transport is an important phenomenon in this case, since L_L is quite large.

The boundary layer concept and Equation (2.17) can also be applied to model mass transfer between other compartments, e.g., the exchange of oxygen between the bulk liquid and the atmosphere. Furthermore, Equation (2.17) can be used to model mass transfer of dissolved components through permeable membranes. Alternative mathematical models for mass transfer through boundary layers can be found in Fogler (1999).

The exchange of particulate components between the biofilm solid matrix and the bulk liquid is modeled by the processes detachment and attachment. Detachment is the transfer of cells and other particulate components from the biofilm solid matrix to the bulk liquid. If large chunks of biomass are instantaneously disrupted from the solid matrix, the process is referred to as sloughing. If detachment involves only single cells or relatively small particles, it is called erosion. Attachment is the deposition and adhesion of cells and other particles suspended in the bulk liquid at the biofilm solid matrix.

The net interfacial transfer rate, r_A ($ML^{-2}T^{-1}$), depends on a balance of attachment and detachment. Attachment and detachment rates can be described simply as first order in the concentration of attached biomass and the surface accumulation of biofilm, respectively:

$$r_A = k_{at} X_B / a - k_{de} X_F L_F \tag{2.20}$$

where X_B is the concentration of suspended component in the bulk liquid ($M_X L^{-3}$), k_{at} is the attachment rate coefficient (T^{-1}), a is the specific surface area of the biofilm ($L^2 L^{-3}$), $X_F L_F$ is the amount of the component in the biofilm per unit surface area ($M_X L^{-2}$), and k_{de} is the detachment rate coefficient (T^{-1}). Because attached species do not exist in the bulk liquid, concentration continuity is not relevant at the interface. However, continuity is ensured by a flux boundary condition equation analogous to Equation (2.17):

$$r_A = j_n \tag{2.21}$$

In many models, attachment is not explicitly included. In some cases, this is a good simplification because detachment is the dominant process. When this is not the case, however, the simulated detachment process must represent the net transfer of particulate mass from the biofilm to the bulk liquid, implicitly including attachment and detachment processes. General detachment functions such as (2.20) have been used also in multidimensional biofilm models, with the advantage of computational simplicity and the possibility to describe detachment by both erosion and sloughing (Xavier et al. 2005a, 2005b, 2005c).

Detachment phenomena can be considerably more complex than is represented by first-order loss, and recent models attempt to capture some of the complexity. For example, detachment can be described by stochastic events related to the distance that a particulate component is from the substratum (Hermanowicz 1998), to the creation of a discontinuity in the biofilm's solid phase due to biomass decay (Pizarro *et al.* 2001), or to a decay process mediated by a chemical signal (Hunt *et al.* 2003). In an alternative model based on structural mechanics computations, the rupture of the biofilm solid matrix occurs when shear forces induced by liquid flow create stresses in the biofilm exceeding the adhesive strength of the matrix (Picioreanu *et al.* 2001; Laspidou and Rittmann 2005).

2.5   MODEL PARAMETERS

Parameters are intrinsic to a model and their values affect the results as much as the choice of model structure, components, and processes. Parameters can be classified as system specific or universal. System-specific parameters depend on the biofilm system considered; examples are biofilm thickness and density. Universal parameters, such as the diffusion coefficient of a component in pure water or the true yield of a certain biomass-substrate combination, don't change with the biofilm system and they can be obtained from the literature or from experiments carried out independently from the biofilm being modeled.

Parameter values originating from different sources may vary considerably. Reasons for this variability include different definitions of the parameters, different units, or values obtained under different environmental conditions. The following sections discuss the factors affecting the values and significance of model parameters.

2.5.1 Significance of model-parameter definitions

Defining a parameter is intrinsically an assumption underlying the modeled process, and the definition (along with the units) can significantly affect the numerical value of the parameter. If a parameter value is taken from the literature, it is of utmost importance to know how that parameter was originally defined, which conditions prevailed, and which assumptions were made when it was determined.

In most biofilm models, the concentration of a dissolved component is defined as mass per unit biofilm volume. Such a definition implies that the component is evenly distributed between the solid and liquid phases of the biofilm compartment and, therefore, that the whole biofilm volume is available for diffusive transport. Alternatively, Wanner and Reichert (1996) define the concentration of a dissolved component as mass per unit volume of only the biofilm's liquid phase and require that diffusion take place only in the liquid phase of the biofilm compartment. These different definitions mean that the numerical values of the diffusion coefficients in the biofilm compartment, D_F, used in each type of model are different and cannot be interchanged. Because Wanner and Reichert (1996) assume that diffusion only takes place in the liquid phase, their D_F is equivalent to the diffusion coefficient of the component in pure water D_L. In contrast, when diffusion is assumed to occur over the entire biofilm compartment, the diffusion coefficient is typically calculated as $D_F = fD_L$, where f is a correction factor (commonly about 0.8, but ranging from about 0.2 to about 1) that accounts for additional mass transport limitations caused by the presence of the solid phase. Wanner and Reichert (1996) defined the term ε_l as the ratio of liquid phase volume to total biofilm volume. If diffusion occurs mainly in the biofilm's liquid phase, ε_l is equals f. However, f can be larger or smaller than ε_l if other factors affect the diffusion of a solute. For example, f can be larger than ε_l if the solute readily diffuses through biomass, but smaller than ε_l if the solute adsorbs to the biomass.

Kinetic parameters for microbial reactions offer more examples of universal parameters that are available in the literature, but that can have quite different definitions. Values for microbial kinetic parameters typically are determined experimentally in suspended cultures under the assumptions of no interactions between individual suspended cells and no spatial gradients in the system. However, when the experiments are performed with microbial flocs, such as occur with activated sludge, concentration gradients and mass-transfer limitations violating the underlying assumptions. For example, an experimental value obtained for the Monod half-maximum-rate concentration K_S (see Equation 2.6) is an "apparent value" or "system-specific value" that is higher than the true value. Beccari *et al.* (1992) demonstrated that much of the variability of the published values of K_S can be explained by the variable effects of mass-transfer limitation. Since mass transfer resistance is explicitly taken into account in biofilm models, true (intrinsic) values of this parameter must be used in the models, not apparent values. The problem of apparent versus true values also is acute for the yield coefficient, in which decay give an apparent yield value that is smaller than true Y value.

A final example of differing definition concerns the detachment/decay rate coefficient. Biofilms do not grow infinitely because loss processes, such as endogenous decay and the detachment of biomass from the biofilm, balance synthesis. In continuum models, detachment is generally described as a first-order process with a detachment rate coefficient k_{de}, as in Equation 2.18. The process can be assumed to occur only at the biofilm surface, equally over the entire biofilm depth, or with different rates at different locations (Reichert and Wanner 1997). This assumption significantly affects the distribution of biomass types from a multi-species biofilm model, a feature that is illustrated in BM3 (Chapter 4). If k_{de} describes detachment equally over the entire biofilm depth, its value can be lumped with of other loss processes, such as lysis, endogenous respiration, and protozoan grazing. Lumping the loss processes simplifies the mathematical solution, but may be misleading when the loss processes have different rate laws and products (e.g., detached biomass versus inorganic carbon for endogenous respiration). Therefore, the processes that are included in the value of k_{de} should always be clearly defined,

2.5.2 Significance of model parameter units

In modeling, it is of utmost importance that the units of variables and parameters are consistent. Thus, a first step in developing or applying a model is to define appropriate units for the components, processes, and parameters that are included in the model. Especially for parameter values taken from the literature, the units have to be checked carefully, and, if needed, the values have to be converted to the common set of units used in the model.

For instance, the concentration of electron donor substrate can be expressed in mol/L, C-mol/L, g/L, gC/L, or based on chemical oxygen demand (COD) as gCOD/L. An electron-donor substrate defined as a distinct chemical compound can be measured in mol/L or g/L, but aggregated donors are measured by lumped parameters like gCOD/L or gC/L. Stoichiometry and the oxidation state (OS) of the carbon of an organic donor can be used to convert between systems of units (Rittmann and McCarty 2001). For example, 1 gCOD/L = (0.67(4-OS)) gC/L, where OS is the oxidation state of the carbon in the organic substrate. For glucose, OS = 0, and the conversion is 1 gCOD/L = 2.67 gC/L. However, a more reduced compound like benzene (C_6H_6) has OS = -1, and its conversion is 1 gCOD/L = 3.33 gC/L. An oxidized compound like formic acid (CH_2O_2) has OS $\sim$ +2, making the conversion 1 gCOD/L $\sim$ 1.33 gC/L.

Microbial mass is another parameter that can be expressed in different units, such as mol, C-mol, g dry weight (DW), g volatile solids (VS), g C, g COD, g protein, or cell numbers. The choice of units in a model sometimes depends on how the different components are measured experimentally. Analyses of volatile solids, dry weight, and suspended COD are simple and convenient ways to measure microbial mass. The typical relationships among them are 1 g VS = 1.1 g DW = 1.42 g suspended COD.

The different units also represent different characteristics of the dissolved or particulate components. COD represents electron equivalents (e⁻ eq), with each e⁻ eq equal to 8 gCOD. Oxygen itself can accept electrons in the same proportion, 1 e⁻ eq per 8 gO$_2$. It is convenient to represent dissolved and particulate components using COD, because then all components are tracked by the same characteristic. C represents the carbon content, and each C equivalent has 12 gC. VS are the organic or combustible materials in cells and organic particles and typically include a fraction of C of about 50%. Another major component of biomass is protein, which also accounts for about 50% of the VS. Cell numbers often are the most difficult units for expressing the mass of microorganisms, because microbial cells vary in size. The conversion of cell numbers to mass units, like gVS, depends on the cell volume. If V_{cell} is the volume of one cell expressed in µm^3 and n is the number of cells per liter, 1 gVS/L = $10^{-12} \times n \times V_{cell}$.

Other parameter units must be consistent with the units of the dissolved and particulate components. For instance, if microbial mass is expressed as gVS and substrate mass as gCOD, then the true yield coefficient must be in gVS/gCOD. If a yield coefficient value found in the literature is expressed in g protein/gCOD, its numerical value will be very different from its numerical value in gVS/gCOD. Since the dimension of the yield coefficient is mass per unit mass, this problem can easily be overlooked, but changing units without converting the numerical value can lead to large errors.

2.5.3 Significance of environmental conditions

All quantities that are not explicitly considered in the model as variables or parameters are referred to as environmental conditions. For instance, if alkalinity (i.e., HCO$_3^-$) is not explicitly calculated by the model, it represents an environmental condition, even if HCO$_3^-$ is found within the modeled system and the modeled processes are affected by its concentration. The justification to treat HCO$_3^-$ as an environmental condition originates from the assumption that the concentration of HCO$_3^-$ does not significantly vary with time or space in the considered biofilm system. Based on this assumption, the values of the parameters that depend on alkalinity can be adjusted according to the actual concentration of HCO$_3^-$. If this adjustment is not made, the model will produce incorrect results, as is demonstrated below for an example of autotrophic growth.

In the ASM activated sludge model Nr. 3 - ASM3 (Henze *et al.* 2000) - the specific growth rate of the autotrophic organisms, μ_A (T^{-1}), is modeled as

$$\mu_A = \mu_{max,A,10} e^{0.1(T-10)} \frac{S_{HCO_3^-}}{K_{HCO_3^-} + S_{HCO_3^-}} \frac{S_{O_2}}{K_{O_2} + S_{O_2}} \frac{S_{NH_4^+}}{K_{NH_4^+} + S_{NH_4^+}} \quad (2.22)$$

where $\mu_{max,A,10}$ is the maximum specific growth rate at 10 °C (T^{-1}), T is the temperature (θ in °C), S is the concentration of a soluble component (ML^{-3}), K is a half saturation constant (ML^{-3}), and O$_2$, HCO$_3^-$ and NH$_4^+$ are subscripts denoting dissolved oxygen, alkalinity, and ammonium, respectively. A useful simplification of Equation (2.22) is the following:

$$\mu_A = \mu_{A,max} \frac{S_{O_2}}{K_{O_2} + S_{O_2}} \frac{S_{NH_4^+}}{K_{NH_4^+} + S_{NH_4^+}} \tag{2.23}$$

Comparison of Equations (2.22) and (2.23) reveals that the two rate equations are likely to produce different results if $\mu_{A,max}$ is given the same numerical value as $\mu_{A,max,10}$. On the other hand, the discrepancy disappears when $\mu_{A,max}$ is calculated with the actual temperature and the actual concentration of HCO_3^- as

$$\mu_{A,max} = \mu_{A,max,10} e^{0.1(T-10)} \frac{S_{HCO_3^-}}{K_{HCO_3^-} + S_{HCO_3^-}} \tag{2.24}$$

This example demonstrates that environmental conditions can affect the values of model parameters and illustrates a proper way to correct parameter values according to these environmental conditions. Environmental conditions that most often affect parameter values are temperature, pH, alkalinity and nutrient concentration.

2.5.4 Plausibility of parameter values

Because the values of model parameters can significantly affect modeling results, the plausibility of these values should be checked either experimentally or by theoretical considerations. In some cases, the values of model parameters have theoretical bounds that must not be violated.

A good example is the true yield coefficient, Y ($M_X M_S^{-1}$), which describes the fraction of substrate that is used for cell synthesis. Since the amount of substrate converted to microbial mass cannot be larger than the amount of substrate utilized, an upper bound for the value of the yield coefficient is established. However, it should be noted that the value of this upper bound depends on the units used and is different if expressed as gCOD/gCOD, gVS/gCOD, or g protein/g substrate. When the units are gCOD/gCOD, the value of the yield cannot exceed the maximum fraction $f_S°$ of electrons used for biomass synthesis reactions per electrons provided by the donor substrate (Rittmann and McCarty 2001). In aerobic systems, $f_S°$ is 0.6-0.7 for heterotrophic organisms and about 0.10-0.15 for nitrifying organisms. Reasonable $f_S°$ values for other microbial metabolisms can be found in Rittmann and McCarty (2001).

There are other plausibility checks: For each transformation process, the stoichiometric coefficients v_{ij} of all components affected by a process must fulfill the condition posed by Equation (2.14). Parameters describing microbial kinetics are well studied, and their values should stay within the bounds considered realistic for each group of microorganisms. For example, the value of the diffusion coefficient in the biofilm compartment is related to the pure-water value, which are readily available (e.g., CRC 1992).

2.5.5 Sensitivity of model parameters

In some cases, a large change in the value of a parameter has no impact on the model output of interest. In other cases, a small change in a parameter value has a profound impact. Sensitivity analysis is used to determine whether or not the choice of a parameter's value has a large, small, or no impact on the model outputs of interest. For given conditions, the number of model parameters that have a significant influence on a model output often is small. Sensitivity analysis helps the modeler decide which parameters must have highly accurate values, while other parameters only need to be in the "right range."

Several computer programs are designed to perform sensitivity analysis. AQUASIM (Reichert 1994, 1998a, 1998b), which is widely used to simulate 1d multi-species and multi-substrate biofilms (Wanner and Reichert 1996), has built-in sensitivity functions. These functions calculate the relative or absolute change of each variable per unit change of any parameter value. When built-in functions are not available, the modeler can accomplish the same goal by first establish reasonable estimates of all parameter values for the conditions of the simulation. Then, one model parameter is given a different value, and the effect of this change is calculated for all outputs of interest. The difference in outputs reflects the effect of the parameter, which should be compared with the uncertainty attributed to the value assigned to the parameter. The sensitivities of different parameters can be compared by performing this procedure for all model parameters.

2.5.6 System-specific parameters

System-specific parameters are particular to a certain biofilm system. Examples come from geometry, hydrodynamics, and the characteristics of the biofilm solid phase. Geometric system parameters are, for example, the bulk liquid volume, the biofilm surface area, and the biofilm thickness. Examples of hydrodynamic parameters are the mean flow velocity, the coefficient of turbulent dispersion, and the concentration boundary layer thickness. Biofilm system parameters include the density of the biofilm solid matrix and the quantities that characterize detachment. The values of system-specific parameters cannot be simply adopted from another system or from the literature, but have to be estimated according for the particular biofilm system being analyzed.

If possible, the values of system-specific parameters should be measured directly. However, direct measurements are sometimes difficult and often involve considerable effort. Therefore, it is a good investment to determine which system-specific parameters do not have a significant effect on the modeling results and are able to be estimated within a reasonable range.

Some system-specific parameters are most easily estimated by an indirect measurement, not directly on the biofilm itself, for instance, the detachment coefficient k_{de} (LT^{-1}) can be determined based on a mass balance of particulate material collected in the system influent and effluent of a biofilm reactor:

$$A_F k_{de} X_F L_F = Q(X_{ef} - X_{in}) \qquad (2.25)$$

where X_{in} and X_{ef} are the concentrations of particulate material in the system influent and effluent ($M_X L^{-3}$), respectively; X_F is the biomass per unit volume biofilm ($M_X L^{-3}$); L_F biofilm thickness (L); Q is the volumetric flow rate ($L^3 T^{-1}$), and A_F is the biofilm surface area (L^2). The value of k_{de} is computed by substituting the measured value for all other terms.

2.6   GUIDANCE FOR MODEL SELECTION

Because most biofilms are very complex systems, a biofilm model that attempts to capture all the complexity would need to include (i) mass balance equations for all processes occurring for all components in all compartments, (ii) continuity and momentum equations for the fluid in all compartments, and (iii) defined conditions for all variables at all system boundaries. Implementing such a model is impractical, maybe impossible. Even the most complex biofilm models existing today contain many simplifying assumptions. Most biofilm models today capture only a small fraction of the total complexity of a biofilm system, but they are highly useful. Thus, simplifications are necessary and a natural part of modeling. In fact, the "golden rule" of modeling is that *a model should be as simple as possible, and only as complex as needed.*

Good simplifying assumptions are identified by a careful analysis of the characteristics of a specific system. These good assumptions become part of the model structure; in other words, they serve as guidance for the selection of the model. The models found in the literature can be differentiated by their assumptions, which depend on the objectives of the modeling effort and the desired type of modeling output. Thus, a user that is searching for a model to simulate specific features of a biofilm system should begin by evaluating the type of assumptions used in creating the models.

One of the objectives of the IWA Task Group on Biofilm Modeling was a comparison of characteristic biofilm models using benchmark problems. A main purpose was to analyze the significance of simplifying assumptions as a prelude for providing guidance on how to select a model. Chapter 3 of this report describes the models used in the benchmark tests, including their relevant features and some guidance for their use. Chapter 4 presents the results of the benchmark problems. The results obtained from the benchmark tests, along with the collective experience of the members of the Task Group, are the basis for the general guidelines for model selection offered in this section.

2.6.1 Overview of the models

The models used by the Task Group can be grouped into four distinct categories according to the level of simplifying assumptions used: namely, analytical (A), pseudo-analytical (PA), 1d numerical (N1), and 2d/3d numerical (N2/N3). As a baseline, all model types normally can represent biofilms having the following features: (i) the biofilm compartment is homogeneous, with fixed thickness and attached to an impermeable flat surface, (ii) only one substrate limits the growth kinetics, (iii) only one microbial species is active, (iv) the bulk liquid compartment is completely mixed, and (v) the external resistance to mass transfer of dissolved components is represented with a boundary layer compartment with a fixed thickness.

Table 2.3. Features by which various types of models of biofilm system differ

Feature	A	PA	N1	N2/N3
Development over time (i.e., dynamic)	−	−	+	+
Heterogeneous biofilm structure	−	−	o[1]	+
Multiple substrates	o[2]	o[2]	+	+
Multiple microbial species	o[3]	o[3]	+	+
External mass transfer limitation predicted	o[4]	o[4]	+	+
Hydrodynamics computed	−[5]	−[5]	−[5]	+[6]

[1] Gradients of all properties possible, but in directions perpendicular to the substratum only
[2] Maximum specific utilization rate of substrate must be adjusted. Concentrations of substrates are based on stoichiometry of biochemical reactions
[3] For steady state modeling an *a priori* distribution must be imposed
[4] No attachment of cells and particles at the biofilm surface
[5] The plug flow can be approximated by a series of completely mixed bulk liquid compartments
[6] Whereas 2d hydrodynamic and mass transfer calculations are very fast on ordinary computers, 3d hydrodynamics is just beginning now to be affordable to common users.

Table 2.3 identifies other features that can be incorporated into certain models and that differentiate among the model types. A plus sign (+) means that the feature can be simulated, a minus sign (−) indicates that the model cannot simulate that feature, and a zero (o) indicates that the model may be able to simulate the feature, but with restrictions that are specified in a footnote. In general, the flexibility and complexity of the models is lower on the left hand side of the table and increases towards the right hand side.

2.6.2 Modeling objectives and user capability

Selecting a model is intimately related to the modeling objectives and the modeling capability of the user of the model. Common quantitative objectives are the calculation of substrate removal, biomass production and detachment rates, or the quantity of biomass present in a given biofilm system. For existing systems, these quantities can also be determined experimentally, but for hypothetical systems, the biofilm model is used strictly as a predictive tool. In engineering applications, biofilm models also are employed to optimize the operation of existing biofilm reactors and to design new reactors. In research, they serve as tools to fill gaps in our knowledge, as they help to identify unknown processes and to provide insight into the mechanisms of these processes. The capability of the user relates to the computing power available and, equally important, to the user's capacity for understanding the model. A model that cannot be formulated or solved by the user is of no value, whether or not it addresses the objectives well.

Simplifying assumptions are related to making the modeling objective mesh with the user capability. For instance, if the objective is to describe the performance of a biofilm system at the macroscale, then the various compartments and processes do not need to be described in too much of a microscale. A lot of microscale detail makes the model difficult to create and computing-intensive. For example, a 1d model with only one type of active biomass may be completely adequate to estimate the flux of one substrate averaged over square meters.

When an existing biofilm system is modeled, known conditions in that system can be used to determine what simplifying assumptions are reasonable. For example, if the biofilm is deep, then the biofilm thickness of the biofilm solid matrix is not a model feature of importance when substrate flux is the objective, since thickness does not affect flux for a deep biofilm. Likewise, if a model is developed to be a design tool for biofilm reactors, then the same features that are anticipated for the biofilm reactor (e.g., mass-transport resistance and specific surface area) must be matched in the modeling process.

If the objective is to model micro-scale processes (e.g., the interaction between microbial cells and EPS in the biofilm or 3d physical structures at the µm scale), the number and type of processes occurring in each compartment of the biofilm need to be represented in microscale detail. For example, a 2d or 3d model is necessary if understanding the physical structure of the biofilm at the µm-scale is the modeling objective, while a multi-species model is necessary if the objective is to understand how ecological diversity develops. Competition of particulate components for space does not only involve metabolically active species. Often, large portions of the biofilm consist of inert, metabolically inactive particulate components, which need to be included in multi-species models. When microscale detail is required, the size of the system being modeled will need to be small in order to make the model's solution possible.

Although many processes always take place in a biofilm, it is not necessary to include every one, depending on the objectives. For example, the spatial distribution of the particulate components can be specified by an *a priori* assumption, instead of being predicted by the model, if the goal is to predict substrate flux for a known biofilm. Then, the model need not include the processes of microbial growth and loss. On the other hand, when the objective is to predict the distribution of microbial species within the biofilm or to calculate the expected biofilm thickness at steady state, then microbial growth and detachment processes are essential.

The transport of dissolved components in the biofilm liquid phase is due to processes beyond molecular diffusion (advection and turbulent dispersion) in some cases. 2d and 3d models, but not 1d models, make it possible to include advection and dispersion by calculating the flow field in the bulk liquid and in the biofilm.

First-order and zero-order rate expressions are common simplifications for the Monod rate of substrate utilization and biomass synthesis. Using these simpler expressions can simplify the mathematical treatment so that the mass balance equations can be solved analytically. However, the simplified rate expressions are valid only for high or low substrate concentrations compared to K_S, and the substrate concentration should be in the proper range throughout the biofilm.

Biofilms do not grow infinitely, and models that consider biofilm accumulation over the long term must incorporate processes by which microbial synthesis is balanced by loss processes. The loss processes can include detachment of cells and particles from the biofilm solid matrix, decay and lysis of cells, hydrolysis of particles, endogenous respiration and maintenance. Thus, when the accumulation of biofilm is a modeling objective, the most important of these loss processes must be identified and included.

Confocal laser scanning microscopy has revealed a large variety of matrix structures, ranging from dense layers of microbial cells (Figure 2.7C) to "mushroom" type biofilms with large channels and pores between clusters of cells (Figure 2.7A). To describe the development of complex structures, the model must represent the interactions of biomass synthesis and decay mechanisms over time and in 2 or 3 dimensions.

A tool to gain insight into whether or not complex structures are likely to be important is the dimensionless G-factor (Picioreanu *et al.* 1998a):

$$G = \frac{L_F^2 \mu_{max} X_F}{DS_B} \qquad (2.26)$$

where X_F is the concentration of bacteria in the biofilm and S_B the concentration of the growth rate-limiting substrate in the bulk liquid.

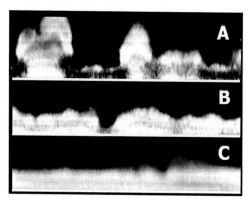

Figure 2.7. Depending on growth conditions, *Pseudomonas aeruginosa* biofilms can have structures ranging from simple flat (C) to very irregular (A). Figure based on CLSM pictures kindly provided by Søren Molin, Center for Biomedical Microbiology, Technical University of Denmark, Lyngby.

Based on modeling experience, the biofilm is likely to have a dense solid matrix and a flat surface when $G < 5$. If $G > 10$ the biofilm is likely to develop complex structures, such as the "mushroom" clusters and channels shown in Figure 2.7A. In the latter case, the user will have to balance the benefits of simulating the structure of the biofilm matrix with the added computational complexity of a multi-dimensional model.

2.6.3 Time scale

Figure 2.8 shows the typical ranges of characteristic times for processes that take place in a biofilm system. The characteristic times differ by as many as 10 orders of magnitude, which offers the possibility to reduce drastically the computational effort without sacrificing accuracy! Usually, our interest in biofilm behavior is narrowed to an "observation window," a certain time scale that can vary from minutes to months, depending on the processes of most interest. For example, biofilm structure has a time scale $> 10^5$ sec (~1 day), while hydrodynamic conditions change with a time scale usually less than 1 sec. Processes that are much faster than our observation window can be described by a pseudo steady-state approximation. Processes that are slow in comparison to the length of the observation window can be described as if they were "frozen" in time.

The order of magnitude of the characteristic times can be established from non-dimensional forms of model equations (Kissel et $al.$ 1984; Picioreanu et $al.$ 2000b). For example, the characteristic time for biomass growth τ_{gr} is usually much larger than the one for substrate transport by diffusion τ_{dif}:

$$\tau_{gr} = \frac{1}{\mu} \gg \tau_{dif} = \frac{L_F^2}{D} \qquad (2.27)$$

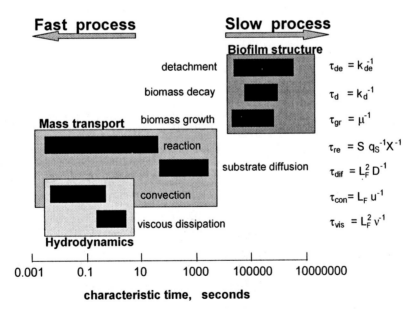

Figure 2.8. Characteristic times of processes occurring in biofilm systems. From Picioreanu et $al.$ (2000b).

This means that, when the bulk liquid substrate concentration suddenly changes, the biofilm solid matrix can be assumed unchanged (i.e., "frozen" state). For instance, biofilm thickness (L_F) and microbial composition and density (X_F) can be assumed constant for the time needed to establish a new steady state profile for substrate. In practical terms, the biomass processes do not need to be modeled to estimate the new substrate flux. Conversely, changes in biofilm structure occur over days or weeks, and the profiles of the dissolved components can be calculated by a steady-state model, while the biofilm itself is modeled as dynamic. After some time, when the shape and composition of the solid matrix have changed enough, the steady state distribution of the dissolved components has to be recalculated. This concept is applied in most models (e.g., Rittmann and Brunner, 1984; Picioreanu et al. 1998, 2000b, 2001, 2004; Eberl et al. 2000a, 2001; Pizarro et al. 2001; Laspidou and Rittmann 2004) that simulate the development in time of the biofilm solid matrix and of the particulate components by which the matrix is formed. Hydrodynamic calculations stabilize so rapidly that the other biofilm components can be put into a frozen state; then, after the hydrodynamic calculations are completed, the state of the other components can be recalculated.

Computer software (e.g., Gear 1971; Wanner and Reichert 1996) is available to handle solutions when processes have very different time scales; the terminology for such a situation is solving a set of stiff differential equations. The advantage of using a stiff solver is that simulations for short and long times can be done with the same model and without making *a priori* decisions about frozen state and steady state. The trade-off is that the stiff solver has a high computational load.

2.6.4 Macro versus micro scales

Biofilm models can be used to provide information at macro-scale or micro-scale. *Macro-scale* outputs include substrate removal rates, biomass accumulation in the biofilm and biomass loss from the system. Typical *micro-scale* outputs are the spatial distributions of substrates and microbial species in the biofilm. In this report, the macro- and micro-scale model outputs are used as outputs to compare the performance of the range of models used in the benchmark tests (Chapter 4). We discuss how the models can be used to find typical macro-scale and micro-scale outputs.

2.6.4.1 Substrate removal

Substrate removal is a main macro-scale output, particularly in the design and characterization of biofilm reactors used in engineering applications. Substrate removal can be expressed in different forms, but each is derived from a global mass balance over the whole biofilm system, as illustrated in Figure 2.9. For steady state (no change in accumulation) and with only bulk liquid and biofilm compartments:

$$Q_{in}S_{in} - Q_{ef}S_{ef} - F_{B,S} - F_{F,S} = 0 \qquad (2.28)$$

where Q_{in} and Q_{ef} are the influent an effluent volumetric flow rates (L^3T^{-1}), respectively, S_{in} and S_{ef} are the influent an effluent substrate concentrations (M_SL^{-3}), respectively, and $F_{B,S}$ and $F_{F,S}$ are the overall substrate consumption rates in the bulk liquid and biofilm (M_ST^{-1}), respectively (Figure 2.9).

The substrate-related outputs usually of interest can be related directly to the terms in Eq. (2.28):
(1) the substrate concentration in the effluent, S_{ef} (M_SL^{-3})

(2) the substrate mass removed per unit time in the whole system, (M_ST^{-1})

$$F_{rem,S} = Q_{in}S_{in} - Q_{ef}S_{ef} = F_{B,S} + F_{F,S} \qquad (2.29)$$

(3) the degree of substrate conversion, which is the fraction of substrate mass removed per unit influent mass (-),

$$x_S = \frac{F_{rem,S}}{Q_{in}S_{in}}$$

(2.30)

(4) the rate of substrate removal per unit system volume ($M_S T^{-1} L^{-3}$)

$$r_{rem,S} = \frac{F_{rem,S}}{V}$$

(2.31)

(5) the rate of substrate removal per unit biofilm area ($M_S T^{-1} L^{-2}$)

$$r_{rem,A,S} = \frac{F_{rem,S}}{A_F}$$

(2.32)

where V is the system volume (L^3) and A_F is the biofilm area (L^2). [The biofilm area (A_F) is considered here being equivalent to the area of substratum (A_S) on which the biofilm grows.] If no substrate conversion takes place in the bulk liquid, the rate of substrate removal per unit biofilm area is the flux of substrate entering the biofilm ($M_S T^{-1} L^{-2}$):

$$r_{rem,A,S} = \frac{F_{F,S}}{A_F} = j_{F,S}$$

(2.33)

where $j_{F,S}$ is the overall flux of substrate through the biofilm surface ($M_S L^{-2} T^{-1}$).

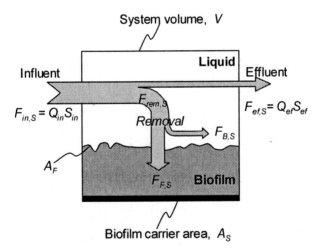

Figure 2.9. Steady state substrate mass balance for a biofilm system including a bulk liquid compartment and the biofilm. Substrate flows: $F_{in,S}$ - substrate in influent, $F_{ef,S}$ - substrate in effluent, $F_{B,S}$ - substrate removed in bulk liquid, $F_{F,S}$ - substrate removed in biofilm, $F_{rem,S}$ - global substrate removal.

2.6.4.2 Biomass accumulation, production, and loss

Biomass accumulation, production, and loss are key macro-scale outputs for the particulate components of a biofilm. A dynamic (nonsteady state) mass balance on the total biomass accumulated in the system, F_X (M_XT^{-1}) – including mass $F_{F,X}$ in the biofilm ("fixed" biomass) and mass $F_{B,X}$ suspended in the bulk liquid ("suspended" biomass) – is:

$$F_X = F_{F,X} + F_{B,X} = F_{in,X} - F_{ef,X} + F_{B,gro,X} + F_{F,gro,X} \qquad (2.34)$$

where $F_{in,X}$ and $F_{ef,X}$ are total flow rates of biomass entering and leaving the system with the influent and effluent, respectively, and $F_{B,gro,X}$ and $F_{F,gro,X}$ are overall net biomass production rates (growth) in suspension and in the biofilm, respectively (M_XT^{-1}). The rates are illustrated in Figure 2.10. Similar to equation (2.29), expressed in terms of liquid flow rates Q_{in} and Q_{ef}, the overall biomass balance in the system at steady state is:

$$Q_{in}X_{in} - Q_{ef}X_{ef} = F_{B,gro,X} + F_{F,gro,X} \qquad (2.35)$$

where X_{in} and X_{ef} are the biomass concentrations contained in influent and effluent (M_XL^{-3}), respectively.

The steady-state macro-scale outputs concerning biomass production are related directly to terms in equation (2.35):

(1) the biofilm biomass per unit carrier area, $X_{F,A}$, defined for total biomass as well as for individual particulate components, (M_XL^{-2})

(2) the biomass produced per unit time in the whole system (M_XT^{-1})

$$F_{pro,X} = Q_{in}X_{in} - Q_{ef}X_{ef} = F_{B,gro,X} + F_{F,gro,X} \qquad (2.36)$$

(3) the global (i.e., system-averaged) rate of biomass production per unit system volume ($M_XL^{-3}T^{-1}$)

$$r_{pro,X} = \frac{F_{pro,X}}{V} \qquad (2.37)$$

(4) the rate of biomass production per unit biofilm area ($M_X L^{-2}T^{-1}$)

$$r_{pro,A,X} = \frac{F_{pro,X}}{A_F} \qquad (2.38)$$

where V is the system volume (L^3) and A_F is the biofilm area (equal to the biofilm substratum area, A_S) (L^2).

(5) the biomass loss (sludge production) per unit time (M_XT^{-1})

$$F_{ef,X} = Q_{ef}X_{ef} = Q_{in}X_{in} + F_{B,gro,X} + F_{F,gro,X} \qquad (2.39)$$

Sludge production is the loss of biomass in the form of suspended cells and flocs in the effluent.

(6) the overall rate of biomass detachment equals the rate of biomass growth in the biofilm with no biomass entering the system ($X_{in}=0$) and negligible attachment, ($M_XL^{-2}T^{-1}$)

$$r_{A,de,X} = \frac{F_{de,X}}{A_F} = \frac{F_{F,gro,X}}{A_F} \qquad (2.40)$$

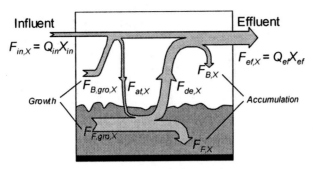

Figure 2.10. Biomass flows in a biofilm system: $F_{in,X}$ - biomass in influent, $F_{ef,X}$ - biomass in effluent (sludge production), $F_{B,X}$ - biomass accumulated in suspension, $F_{F,X}$ - biomass accumulated in biofilm, $F_{F,gro,X}$ - biomass growth in biofilm, $F_{B,gro,X}$ - biomass growth in suspension, $F_{at,X}$ - biomass attached from suspension to biofilm, $F_{de,X}$ - biomass detached from biofilm.

2.6.4.3 Spatial profiles of dissolved components

The profiles of dissolved components are common micro-scale outputs of interest. One-dimensional models can provide *concentration profiles* of dissolved components along the biofilm depth z, i.e. $S(z)$. The multi-dimensional models calculate *concentration fields* as, for example, $S(x,y,z)$ in 3d (see Figure 2.3B). In the benchmarks, other simple micro-scale indicators are used as well, such as the average substrate concentrations at the biofilm surface and at the biofilm base, S_{LF} and S_0, respectively. A comparison of S_{LF} with bulk phase concentrations S_{ef} is used to assess the significance of an external mass transfer resistance. Similarly, the comparison S_0 with S_{LF} is used to evaluate the importance of internal mass transfer resistances and to estimate the substrate penetration in the biofilm.

2.6.4.4 Spatial distribution of particulate components

Different particulate components (e.g., biomass types such as heterotrophs, nitrifiers, and EPS) can have very different spatial distributions in the micro-scale. Therefore, not only the total biomass accumulation, but also the distribution of individual particulate components is a key micro-scale output. Some models compute the biomass distribution in the biofilm, while other models need to have the distribution provided as an input. Some 1d biofilm models calculate biomass concentration profiles along the biofilm depth z, i.e. $X_F(t,z)$. Multidimensional biofilms models compute the dynamics of the 2d and 3d biomass distribution as, for example, $X_F(t,x,y,z)$ in 3d (Figure 2.4 and Figure 2.5).

2.6.4.5 Physical structure of the biofilm

The shape and local density of the biofilm solid matrix is a micro-scale output that is growing in importance (Figure 2.7). Models in 2d or 3d are needed to generate these microscale details of physical structure. Some examples of model-generated biofilm structure are presented in Section 3.6.5 and in Figures 2.3-2.5.

3

Biofilm models

3.1 MASS BALANCES IN BIOFILM MODELS

The most basic principle for all quantitative models is conservation of mass. Conservation of mass of a component in a dynamic and open system states that:

$$\begin{pmatrix} \textit{Net rate of} \\ \textbf{\textit{accumulation}} \\ \textit{of mass} \\ \textit{of component} \\ \textit{in the system} \end{pmatrix} = \begin{pmatrix} \textit{Net rate of} \\ \textit{mass influx} \\ \textit{of component} \\ \textbf{\textit{transported}} \\ \textit{into the system} \end{pmatrix} + \begin{pmatrix} \textit{Net rate of} \\ \textbf{\textit{generation}} \\ \textit{of the} \\ \textit{component} \\ \textit{in the system} \end{pmatrix} \qquad (3.1)$$

The net influx of mass is the difference between the mass brought into the system and the mass leaving the system. Similarly, the net mass of a component generated in the system is the difference between production and consumption:

$$\begin{pmatrix} \textit{Net rate of} \\ \textbf{\textit{accumulation}} \\ \textit{of mass} \\ \textit{of component} \\ \textit{in the system} \end{pmatrix} = \begin{pmatrix} \textit{Mass flow} \\ \textit{of the} \\ \textit{component} \\ \textbf{\textit{into}} \\ \textit{the system} \end{pmatrix} - \begin{pmatrix} \textit{Mass flow} \\ \textit{of the} \\ \textit{component} \\ \textbf{\textit{out of}} \\ \textit{the system} \end{pmatrix} + \begin{pmatrix} \textit{Rate of} \\ \textbf{\textit{production}} \\ \textit{of the} \\ \textit{component by} \\ \textit{transformations} \end{pmatrix} - \begin{pmatrix} \textit{Rate of} \\ \textbf{\textit{consumption}} \\ \textit{of the} \\ \textit{component by} \\ \textit{transformations} \end{pmatrix}$$

$$(3.2)$$

3.1.1 Microscopic (local or differential) mass balances

3.1.1.1 General differential mass balances

At the microscopic level, *local* mass balances can be written based on Eq. (3.2). The local mass balances are differential equations that express the variation of concentration of a component in time in a point in space as a result of transport and transformation processes. The local mass balances are the mathematical form of equality, which in a Cartesian space (i.e., with ortho-normal unit vectors), can be written as

$$\frac{\partial C}{\partial t} = -\frac{\partial j_x}{\partial x} - \frac{\partial j_y}{\partial y} - \frac{\partial j_z}{\partial z} + r \qquad (3.3)$$

where t is time (T); x, y and z are spatial coordinates (L); C is the concentration (ML^{-3}); j_x, j_y and j_z, are the components of the mass flux $\mathbf{j}$ (ML^{-2}T^{-1}) along the coordinates; and r is the net production rate (ML^{-3}) of the component. Equation (3.3) is the *equation of continuity for a component*, either soluble or particulate. If the system geometry requires, similar equations can be formulated in other systems of coordinates, e.g. cylindrical of spherical.

Most commonly, the mass flux $\mathbf{j}$ in Eq. (3.2) is comprised of a diffusive flux $\mathbf{j_D}$ and a convective flux (more accurately called advective flux), $\mathbf{j_C}$. Other contributions to the soluble component flux, neglected in what follows, may come from migration in a field of electrical potential (Nernst-Planck), chemokinesis, and chemotaxis (Patak-Keller-Segel), or from considering activity coefficients in solutions of concentrated electrolytes. The diffusion flux in a direction, e.g., the z direction, is given by Fick's first law

$$j_{D,z} = -D\frac{\partial C}{\partial z} \qquad (3.4)$$

where the diffusion coefficient D (L^{2}T^{-1}) may include besides the molecular diffusion also turbulent diffusion. The convective flux is

$$j_{C,z} = u_z C \qquad (3.5)$$

where u_z denotes the z-component of the carrying fluid velocity field.

Often, the reaction rate r depends on the component concentration C, as explained in Section 2.4.1.

Substituting (3.4) and (3.5) into equation (3.3) yields a second-order partial differential equation

$$\frac{\partial C}{\partial t} = -\frac{\partial (u_x C)}{\partial x} - \frac{\partial (u_y C)}{\partial y} - \frac{\partial (u_z C)}{\partial z} + \frac{\partial}{\partial x}\left(D\frac{\partial C}{\partial x}\right) + \frac{\partial}{\partial y}\left(D\frac{\partial C}{\partial y}\right) + \frac{\partial}{\partial z}\left(D\frac{\partial C}{\partial z}\right) + r \quad (3.6)$$

Using the divergence operator $\nabla \equiv \partial/\partial x + \partial/\partial y + \partial/\partial z$, equation (3.6) can be written more compactly as:

$$\frac{\partial C}{\partial t} = -\nabla(\mathbf{u}C) + \nabla(D\nabla C) + r \qquad (3.7)$$

Solving equation (3.6) or (3.7) means finding the field (i.e., spatial distribution) of concentration C at different moments in time. In other words, the solution gives values for the concentration $C(t,x,y,z)$ of a component in each point in space and in time. The complication when using models including convective mass transport is that the fluid velocity field $\mathbf{u}$ must either be known or calculated, which may be a computationally intensive task. In the liquid region, the carrier of the convective transport is the hydrodynamic flow field, governed by the incompressible Navier-Stokes equations describing conservation of momentum and mass of liquid. These equations are:

$$\frac{\partial \mathbf{u}}{\partial t} + \mathbf{u}\nabla\mathbf{u} = -\frac{1}{\rho}\nabla p + \nu\nabla^2\mathbf{u} + \mathbf{g}, \qquad \nabla\mathbf{u} = 0 \qquad (3.8)$$

where $\mathbf{u}$ denotes the flow velocity vector (LT^{-1}), p is the pressure ($ML^{-1}T^{-2}$), and ρ and ν denote constant density (ML^{-3}) and kinematic viscosity (L^2T^{-1}), respectively, and $\mathbf{g}$ is the acceleration (LT^{-2}) produced by a body force (for instance, gravity). At the liquid/biofilm interface, no-slip boundary conditions typically are used for the flow field (i.e., $\mathbf{u} = 0$).

3.1.1.2 Particular forms of differential mass balances

Many simplified versions of the general mass balance equation (3.6) can be relevant, depending on the characteristics of the biofilm system. For example,

Steady state: $\qquad\qquad \dfrac{\partial C}{\partial t} = 0$

Reduced dimension: $\qquad \dfrac{\partial C}{\partial y} = 0$ for 2d, $\quad \dfrac{\partial C}{\partial x} = \dfrac{\partial C}{\partial y} = 0$ for 1d

No convection: $\qquad\quad \mathbf{u} = 0$

No diffusion: $\qquad\qquad D = 0$

No reaction: $\qquad\qquad r = 0$

Some typical examples of differential mass balances in particular cases are:

a. 3d steady state (*in bulk liquid or reactive boundary layer in general*)

$$-\frac{\partial(u_x C)}{\partial x} - \frac{\partial(u_y C)}{\partial y} - \frac{\partial(u_z C)}{\partial z} + \frac{\partial}{\partial x}\left(D\frac{\partial C}{\partial x}\right) + \frac{\partial}{\partial y}\left(D\frac{\partial C}{\partial y}\right) + \frac{\partial}{\partial z}\left(D\frac{\partial C}{\partial z}\right) + r = 0 \,(3.9)$$

b. 3d diffusion plus reaction, steady state (*in biofilm matrix*)

$$\frac{\partial}{\partial x}\left(D\frac{\partial C}{\partial x}\right) + \frac{\partial}{\partial y}\left(D\frac{\partial C}{\partial y}\right) + \frac{\partial}{\partial z}\left(D\frac{\partial C}{\partial z}\right) + r = 0 \qquad (3.10)$$

c. 3d diffusion plus reaction, constant diffusivities, dynamic or steady state (*in biofilm matrix*)

$$\frac{\partial C}{\partial t} = D\left(\frac{\partial^2 C}{\partial x^2} + \frac{\partial^2 C}{\partial y^2} + \frac{\partial^2 C}{\partial z^2}\right) + r \quad (3.11) \qquad D\left(\frac{\partial^2 C}{\partial x^2} + \frac{\partial^2 C}{\partial y^2} + \frac{\partial^2 C}{\partial z^2}\right) + r = 0 \quad (3.12)$$

d. 3d diffusion, constant diffusivities, dynamic or steady state (*in diffusive boundary layer*)

$$\frac{\partial C}{\partial t} = D\left(\frac{\partial^2 C}{\partial x^2} + \frac{\partial^2 C}{\partial y^2} + \frac{\partial^2 C}{\partial z^2}\right) \qquad (3.13) \qquad D\left(\frac{\partial^2 C}{\partial x^2} + \frac{\partial^2 C}{\partial y^2} + \frac{\partial^2 C}{\partial z^2}\right) = 0 \qquad (3.14)$$

e. 1d diffusion or dispersion and convection plus reaction, dynamic or steady state (*in bulk liquid with axial dispersion*)

$$\frac{\partial C}{\partial t} = -u_x\frac{\partial C}{\partial x} + \frac{\partial}{\partial x}\left(D\frac{\partial C}{\partial x}\right) + r \quad (3.15) \qquad -u_x\frac{dC}{dx} + \frac{d}{dx}\left(D\frac{dC}{dx}\right) + r = 0 \qquad (3.16)$$

f. 1d convection plus reaction, dynamic or steady state (*in bulk liquid with plug flow*)

$$\frac{\partial C}{\partial t} = -u_x\frac{\partial C}{\partial x} + r \qquad\qquad (3.17) \qquad -u_x\frac{dC}{dx} + r = 0 \qquad\qquad (3.18)$$

g. 1d diffusion plus reaction, dynamic (Eq. (3.19)) or steady state (Eq. (3.20)) (*in biofilm matrix*)

$$\frac{\partial C}{\partial t} = \frac{\partial}{\partial x}\left(D\frac{\partial C}{\partial x}\right) + r \tag{3.19}$$

$$\frac{d}{dx}\left(D\frac{dC}{dx}\right) + r = 0 \tag{3.20}$$

h. 1d diffusion and reaction with constant diffusion coefficient:

$$D\frac{d^2C}{dx^2} + r = 0 \tag{3.21}$$

A mass-transfer process may be described by solving one of the differential forms of the mass balances by using the appropriate initial or boundary conditions, or both, depending on the existing physical situation.

Initial conditions for dynamic mass transfer processes refer to the values of concentrations C at the start of the time interval of interest (usually at $t = 0$). The concentration may be simply equal to a constant or a more complex distribution function of space variables.

Boundary conditions refer to the values of C existing at specific positions on the boundaries of the system. The boundary conditions most generally encountered include (see also Figure 3.1):

1. The concentration on a surface may be specified

$$C(surface) = C_{LF} \tag{3.22}$$

For example, when solving mass balances in the biofilm, the concentration at the biofilm surface may be assigned (e.g., the value in the bulk liquid if the resistance to mass transfer through the boundary layer is neglected). (Figure 3.1a)

2. The mass flux at a surface may be specified. Cases of engineering interest include the no-flux condition ($j_{surface} = 0$). The mass flux is zero at an impermeable surface (Figure 3.1d), for a purely diffusive flux leads to

$$\frac{\partial C}{\partial z} = 0 \tag{3.23}$$

for example at the biofilm/substratum boundary. This is also a common condition on axi-symmetric boundaries, for example in the center of a spherical catalyst granule. Another common situation is a specified total flux at the inlet of a channel or column,

$$u_x C - D\frac{\partial C}{\partial x} = j_{in} \tag{3.24}$$

In other cases, the value of the mass flux is specified, for example at a diffusive boundary (Figure 3.1e) on a membrane surface at $z=0$:

$$j_{surface} = -D_F\frac{\partial C}{\partial z}\bigg|_{z=0} = j_0 \tag{3.25}$$

In this way, continuity of mass flux can be applied at the biofilm/liquid interface (Figure 3.1b):

$$j_{surface} = -D_L\frac{\partial C}{\partial z}\bigg|_L = -D_F\frac{\partial C}{\partial z}\bigg|_F = j_F \tag{3.26}$$

where indices F and L denote points adjacent to the biofilm surface in the biofilm and in the boundary layer, respectively. The boundary condition can be defined also in terms of mass transfer coefficients when a fluid is flowing over the phase for which the mass balance is written (Figure 3.1c):

$$j_{surface} = -D_F\frac{\partial C}{\partial z}\bigg|_F = k_c\left(C_{LF} - C_B\right) \tag{3.27}$$

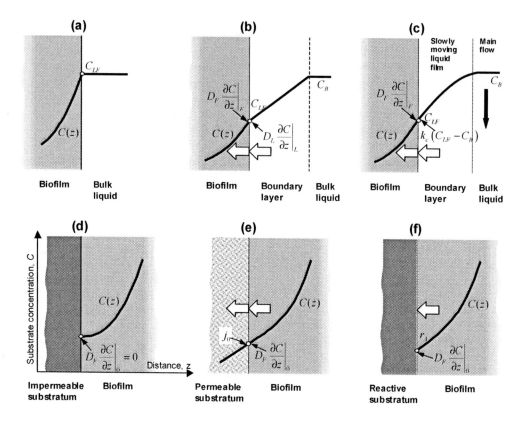

Figure 3.1. Typical boundary conditions encountered in mass transfer processes: (a) specified concentration; (b) continuity of diffusive flux; (c) continuity of total mass flux; (d) no-flux condition – impermeable boundary; (e) specified flux condition – permeable boundary; and (f) reactive boundary.

The rate of a chemical reaction may be specified at a reactive boundary, for example, at the surface of dissolving materials (Figure 3.1f):

$$j_{surface} = -D_F \left. \frac{\partial C}{\partial z} \right|_{z=0} = r_A \qquad (3.28)$$

where the surface reaction rate ($ML^{-2}T^{-1}$) may depend on the concentration of the diffusing species, C.

When the diffusing species disappears at the boundary by an instantaneous reaction, the concentration of that species is often assumed to be zero.

3.1.2 Macroscopic (global or integral) mass balances

3.1.2.1 General integral mass balances

At the macroscopic level, *global* mass balances can be written based on the continuity over the whole biofilm system (Eq. (3.1)). The global mass balances result also from integration of the local balances (Eq. (3.3)) and constitute the main engineering form of mass balance. The global mass balances state that, for any dissolved or particulate component, the change of component mass in time in the system is equal to the difference between component mass flow rate in influent and effluent, plus the net production rate in the system volume. In mathematical terms, for any component, this is written as:

$$\frac{dm}{dt} = F_{in} - F_{ef} + F_{gen} \tag{3.29}$$

where m is the component mass (M), F_{in} and F_{ef} are the component mass flow rates in the influent and the effluent (MT^{-1}), respectively, and F_{gen} is the sum of the rates of all the processes by which the component is produced or consumed (MT^{-1}). For any component, if two compartments, a "bulk liquid" and "biofilm", are distinguished in the system, equation (3.29) for the bulk liquid compartment becomes:

$$\frac{d}{dt}\left(\int_{V_B} C_B dv\right) = F_{in} - F_{ef} + F_F + F_B \tag{3.30}$$

where F_B and F_F are the overall transformation flow rates in the bulk liquid and biofilm (MT^{-1}), respectively, C_B is bulk liquid (and effluent, too, if the compartment is completely mixed) component concentration (ML^{-3}), and finally, V_B is the volume of the bulk liquid compartment (L^3) (Figure 3.2).

In a general case, the influent and effluent mass flow rates result from the integral of fluxes j over the inflow and outflow areas:

$$F_{in} = \int_{A_{in}} j_{in} dA \quad \text{and} \quad F_{ef} = \int_{A_{ef}} j_{ef} dA \tag{3.31}$$

The mass flow rate of component transferred across the biofilm surface, F_F, is expressed by integration of the mass flux j_F (ML^{-2}T^{-1}) across a biofilm surface with area A_F (L^2):

$$F_F = \int_{A_F} j_F dA \tag{3.32}$$

F_F also equals the global transformation of component in the biofilm; thus, an alternative form of equation (3.32) is:

$$F_F = \int_{V_F} r_F dV \tag{3.33}$$

where V_F is the biofilm volume and r_F (ML^{-3}T^{-1}) is a local rate of component transformation within the biofilm volume.

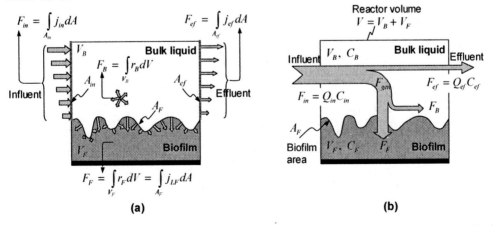

Figure 3.2. Illustration of an integral mass balance and of component mass flow rates in a biofilm system with "bulk liquid" and "biofilm" compartments: (a) Distribution of component fluxes on the system boundaries and rates in the system volume; (b) Global component flow rates.

Similarly, the mass flow rate of a component generated in the bulk liquid, F_B, is the integral of the net generation rate $(ML^{-3}T^{-1})$ of the component, r_B, over the volume (L^3) of the bulk liquid compartment, V_B:

$$F_B = \int_{V_B} r_B dV \qquad (3.34)$$

Substituting all mass flow rates in eq. (3.30) gives a general mass balance in integral form:

$$\frac{d}{dt}\left(\int_{V_B} C_B dv\right) = \int_{A_{in}} j_{in} dA - \int_{A_{ef}} j_{ef} dA + \int_{A_F} j_F dA + \int_{V_B} r_B dV \qquad (3.35)$$

3.1.2.2 Particular forms of the integral mass balance

Usually one knows the influent and effluent volumetric flow rates (L^3T^{-1}), Q_{in} and Q_{ef}, respectively. This leads to

$$F_{in} = Q_{in} C_{in} \qquad \text{and} \qquad F_{ef} = Q_{ef} C_{ef} \qquad (3.36)$$

If it can be assumed that the bulk liquid is perfectly mixed, then the rate r_B is the same in the whole bulk volume and $C_B = C_{ef}$, thus F_B is simply:

$$F_B = V_B r_B \qquad (3.37)$$

Substituting equations (3.36) and (3.37) into (3.35) yields:

$$\frac{d(V_B C_B)}{dt} = Q_{in} C_{in} - Q_{ef} C_B + \int_{A_F} j_F dA + r_B V_B \qquad (3.38)$$

Examples of integral mass balances in some usual particular cases are:

a. Same influent and effluent volumetric flowrates, $Q = Q_{in} = Q_{ef}$, and constant bulk volume V_B:

$$V_B \frac{dC_B}{dt} = Q(C_{in} - C_B) + \int_{A_F} j_F dA + r_B V_B \qquad (3.39)$$

b. Steady state in conditions of point a.:

$$0 = Q(C_{in} - C_B) + \int_{A_F} j_F dA + r_B V_B \qquad (3.40)$$

c. 1d biofilm system means constant flux j_F across the biofilm surface, dynamic (Eq. (3.41)) and at steady state (Eq. (3.42)):

$$V_B \frac{dC_B}{dt} = Q(C_{in} - C_B) + j_F A_F + r_B V_B \qquad (3.41)$$

$$0 = Q(C_{in} - C_B) + j_F A_F + r_B V_B \qquad (3.42)$$

d. No transformations in bulk liquid, in conditions of point c.

$$V_B \frac{dC_B}{dt} = Q(C_{in} - C_B) + j_F A_F \qquad (3.43)$$

$$0 = Q(C_{in} - C_B) + j_F A_F \qquad \text{(BM1, BM2, and BM3)} \qquad (3.44)$$

e. 3d biofilm with no transformations in bulk liquid and steady state:

$$0 = Q(C_{in} - C_B) + \int_{A_F} j_F dA \qquad (3.45)$$

synthesis. The Monod format has the benefit that it naturally describes the gradual transition for first-order kinetics (in substrate concentration) to zero-order (or saturation) kinetics for high substrate concentration. Unfortunately, the non-linear nature of the Monod relationship makes it impossible to find an analytical solution for the substrate mass balance, even when it is the only mass balance to be solved. The simplification that overcomes this problem is to simplify the Monod expression to either zero order or first order in substrate concentration.

Zero-order kinetics can be assumed if the substrate concentration in the bulk liquid is well above the half-saturation concentration (K_S) for the limiting substrate. In cases where the substrate concentration has low values (well below the half-saturation concentration), first-order kinetics can be used. Picking either zero-order or first-order kinetics works well if the substrate concentration is in the high or low range at all points in the biofilm.

When the concentration is not always high or low, a simple compromise approach is to compute the rates by both methods and then compute a weighted average of the zero order and first order rate expressions.

External mass-transport limitation can be included as a boundary condition at the biofilm's outer surface. This boundary condition must be solved simultaneously with the analytical solution for reaction with diffusion inside the biofilm.

It is possible to use analytical solutions for multiple substrates transformed by different microbial groups when the distribution of the biomass species is specified as an input to the model solution. For example, faster growing microorganisms exist at the outside of the biofilm, with slower growing microorganisms near the substratum. Alternately, all the microorganisms can be homogeneously mixed at all positions in the biofilm.

An analytical model is derived for one biofilm compartment with a completely mixed bulk liquid. When the flow pattern of the liquid phase cannot be described with a single well-mixed bulk compartment, multiple compartments can be linked together in series or parallel to achieve a more complex flow regime (Levenspiel 1972).

3.2.2 Definitions and equations

3.2.2.1 Mass balances for substrate in the bulk liquid

The general description of conversion in a completely mixed liquid phase of a biofilm reactor at steady state is based on a mass balance for the component(s) of interest. In analogy with eq. (3.44), for a soluble substrate i consumed in the biofilm the steady-state mass balance is

$$Q\left(S_{im,i} - S_{B,i}\right) = j_{F,i} A_F \tag{3.46}$$

The limiting substrate i can be any soluble component such as organic substrate (S, COD), O_2, NH_3, etc. Equation (3.46) states that the difference between the amount of substrate in the influent and effluent is transferred towards the biofilm, where it is consumed.

3.2.2.2 Mass balances for substrate in the biofilm

In the biofilm, the substrate is transported by diffusion due to the substrate gradient in (only) the x-direction (distance inside biofilm) and is consumed by the biomass present in the biofilm. For a homogeneous flat biofilm in steady state, this can be expressed by eq. (3.21)

$$D_i \frac{d^2 S_{F,i}}{dx^2} + r_i = 0 \tag{3.21}$$

Differential equation (3.21) can be solved analytically only for first- or zero-order reaction kinetics for r_i. For the typical substrate affinity kinetics of biological processes (saturation kinetics) the substrate volumetric conversion rate r_i is:

$$r_i = q_{max,i} \cdot \frac{S_{F,i}}{S_{F,i} + K_i} \cdot X_F \tag{3.47}$$

where X_F is the biomass concentration in the biofilm ($M_X L^{-3}$), $q_{max,i}$ is the maximum specific substrate conversion rate ($M_S M_X^{-1} T^{-1}$) and K_i is the affinity constant ($M_i L^{-3}$) of substrate i. When using (3.47), equation (3.21) can be solved only numerically. However, at high substrate concentration ($S_i \gg K_i$), the reaction can be considered zero order, while at very low concentration ($K_i \gg S_i$) the rate can be considered first order, and in this cases analytical solutions exist.

• *Zero-order kinetics*

The analytical solutions for zero order kinetics differ for fully or partially penetrated biofilms. A penetration depth δ (L) can be defined as the biofilm thickness at which the substrate concentration becomes zero. It is computed from:

$$\delta = \sqrt{\frac{2 \cdot D_i \cdot S_{LF,i}}{q_{max,i} \cdot X_F}} \tag{3.48}$$

where $S_{LF,i}$ is the concentration at the biofilm surface. If no external mass transfer resistance is considered, then $S_{LF,i} = S_{B,i}$.

The biofilm is partially penetrated when it actual thickness is greater than δ. The flux of substrate into a partially penetrated, flat biofilm for zero-order kinetics ($j_{F,i}^{(0)}$) is (Levenspiel 1972; Harremoës 1978; Rittmann and McCarty 2001):

$$j_{F,i}^{(0)} = \sqrt{2 \cdot D_i \cdot q_{max,i} \cdot X_F} \cdot \sqrt{S_{LF,i}} \tag{3.49}$$

where $S_{LF,i}$ is the concentration of substrate i at the liquid-biofilm interface, as external mass transfer is neglected in the applications discussed here (when $S_{LF,i}$ is equal to the concentration in the bulk liquid $S_{B,i}$). The second square root shows that the flux is "half-order" in the substrate concentration. The first square root is the so-called "half-order" rate constant used in the square root approach (Harremoës 1978).

When the biofilm thickness L_F is less than the penetration depth δ, Eq. (3.48) is not appropriate, and the flux must be computed from (Rittmann and McCarty 2001):

$$j_{F,i}^{(0)} = L_F \cdot q_{max,i} \cdot X_F \tag{3.50}$$

• *First-order kinetics*

The flux of substrate into a flat biofilm for first-order kinetics ($j_{F,i}^{(1)}$) is (Levenspiel 1972; Harris and Hansford 1978; Harremoës 1978; Rittmann and McCarty 2001):

$$j_{F,i}^{(1)} = \frac{q_{max,i} \cdot X_F \cdot L_F \cdot S_{LF,i}}{K_i} \cdot \alpha \tag{3.51}$$

where

$$\alpha = \frac{\tanh \beta}{\beta}; \quad \beta = \sqrt{\frac{q_{max,i} \cdot X_F \cdot L_F}{D_i \cdot K_i}} \tag{3.52}$$

where $j_{F,i}^{(1)}$ is the flux of substrate through the biofilm for first-order kinetics ($M_i L^{-2} T^{-1}$) and L_F is the biofilm thickness (L). For a first-order reaction rate, the substrate concentration never becomes zero. Therefore, a penetration depth for the limiting substrate cannot be mathematically defined.

• *Composite kinetics*

A composite flux of substrate through the biofilm ($j_{F,i}$) can be estimated as the weighted average of the zero- and first-order reaction rates:

$$j_{F,i} = \left(\frac{S_{LF,i}}{S_{LF,i} + K_i} \right) \cdot j_{F,i}^{(0)} + \left(1 - \frac{S_{LF,i}}{S_{LF,i} + K_i} \right) \cdot j_{F,i}^{(1)} \tag{3.53}$$

This approach is used in Chapter 4 and also had been used in other biochemical engineering fields, i.e. immobilized biocatalysts (Kobayashi *et al.* 1976; Yamane 1981).

3.2.2.3 Mass balances for biomass

The general biomass balance in steady state for a biofilm process can be written as (Rittmann 1982)

$$j_{F,i} \cdot Y_i - b \cdot X_F \cdot L_F = 0 \tag{3.54}$$

where the first term is the production of biomass (Y_i is the yield biomass/substrate, $M_X M_i^{-1}$) and the second term represents the decrease in biomass due to inactivation, endogenous respiration and detachment. Thus, the overall rate decay coefficient b (T^{-1}) can be computed with the following expression

$$b = b_{ina} + b_{res} + b_{de} \tag{3.55}$$

3.2.3 Mathematical treatment

The concentrations in a biofilm reactor can be computed by solving the mass balance equations for the biofilm simultaneously with the mass balance equations for the liquid phase. The fluxes are the links, because the flux into the biofilm equals the flux out of the liquid. For a set of operating conditions and a given system, influent volumetric flow rate (Q), biofilm surface area (A_F), and influent concentration ($S_{in,i}$) usually are known. For known microorganisms, the kinetic constants, yield, and maximum biomass concentration in the biofilm can be estimated or fixed. If the biofilm thickness is fixed, only the substrate concentration is unknown. If the biofilm thickness also must be computed, then the two unknown variables are the outflow concentration of substrate (which represents the conversion in the reactor) and biofilm thickness (L_F). The system of mass balances has to be solved iteratively.

3.2.3.1 One biological conversion process

For biological reactions, several compounds are converted simultaneously (e.g. COD, oxygen, and N-source). The conversion of all these compounds is linked by the stoichiometric equation describing the biological growth process. One of these compounds will be the rate-limiting compound for which the conversion rate has to be calculated as described above, all other components will be converted relative to their stoichiometric factor. The compound with the relatively lowest transport rate will be the rate-limiting compound. The rate-limiting compound will have the lowest outcome for the following relation (Andrews, 1988):

$$\frac{D_i \cdot S_{B,i}}{Y_i} \tag{3.56}$$

After the rate-limiting component is identified, it has to be evaluated whether this compound is fully or partially penetrating. This can be done by calculating the penetration depth for zero-order kinetics.

3.2.3.2 Two or more biological conversion processes & biofilm architecture

If two or more biological reactions occur, an assumption has to be made on the architecture of the biofilm. From experimental and modeling results, some general assumptions can be made on the biomass distribution with biofilm depth (Wanner and Gujer 1986; Van Loosdrecht *et al.* 1995a; Okabe *et al.* 1999). A first ordering of biomass is based on the type electron acceptor or redox potential. At the outside will be the conversion with the highest redox potential (general aerobic oxidation), while in the inside the conditions get more reduced (anoxic, sulfate reducing and methanogenic). Within a redox zone, a further biomass distribution can occur, where faster growing bacteria are generally found at the outside (e.g. aerobic heterotrophs or acidogens), while slower growing bacteria (nitrifiers or methanogens) are more inside the biofilm. This is especially true with large differences in growth rate. When the differences are small (e.g., ammonium oxidation and nitrite oxidation), one has to consider that the bacteria are mixed. In that case the relative composition of the biomass can be obtained from the ratio of the product of yield coefficient and the converted amount.

As a further simplification, we can assume that the three forms of biomass used in the benchmark models (heterotrophs, autotrophs and inerts) are homogeneously distributed in the biofilm. That is the simplest way to treat a multi-species biofilm. Nevertheless, to take into account the main effect of the real (layered) biofilm structure, limiting conditions for a common substrate for the two species (i.e. oxygen) will always be applied first for the species with lowest growth rate (i.e., autotrophs or nitrifiers).

The competition for space in the biofilm has been taken into account considering the following balance for the total biomass density in the biofilm:

$$X_F = X_{F,H} + X_{F,N} + X_{F,I} \tag{3.57}$$

with X_F the total biomass concentration in the biofilm. $X_{F,H}$, $X_{F,N}$ and $X_{F,I}$ are concentrations of heterotrophic, autotrophic and inert biomass, respectively.

3.2.3.3 Kinetics for multiple limiting substrates

Multiplication of Monod terms is a mathematical way to express the presence of multiple limiting substrates, and it has been shown to be physiologically based for acceptor and donor dual limitation (Bae and Rittmann 1996). The main difficulty will be to handle the multiplication of Monod terms for the two substrates. The mathematical expression used to describe the presence of multiple substrates has also been suggested as the minimum of both terms by several authors (for instance in Hunik *et al.* 1994). In the present study, the approach used is similar to the one used in ASM 1 (Henze *et al.* 1987), modifying the maximum specific substrate conversion rate ($q_{max,i}$) with the Monod term (evaluated in the interface liquid – biofilm) corresponding to the non-limiting substrate:

$$r_i = q_{max,i} \cdot \frac{S_{LF,i}}{S_{LF,i} + K_i} \cdot \frac{S_{LF,O_2}}{S_{LF,O_2} + K_{O_2}} \cdot X_F = q_{max,i,mod} \cdot \frac{S_{LF,lim}}{S_{LF,lim} + K_{lim}} \cdot X_F \tag{3.58}$$

where $q_{max,i,mod}$ is the maximum specific substrate conversion rate modified with the non-limiting saturation terms. This leads to an expression equivalent to the normal Monod expression, but with a variable $q_{max,i}$ term.

3.2.3.4 Solving the problem with a simple spreadsheet

To solve the problem by the weighted average approach, a simple spreadsheet is used to obtain the result. The mass balances can be directly introduced, and it is possible to find the steady state solution using either manual iteration or with the aid of an optimization tool

(SOLVER tool in MS Office Excel), which allows a constrained multivariable optimization. A set of guidelines is given below to describe briefly how the calculations were carried out. For details, the Excel file can be downloaded from

http://www.biofilms.bt.tudelft.nl/analyticalSolutionMaterial/index.html

as an attachment to the publication by Perez *et al.* (2005).

In the case of the automatic iterative process with the optimization tool, for instance, once a set of parameters and conditions is fixed, the target cell is one of the mass balances. The variables for the optimization method (adjustable cells) are the concentration of rate-limiting substrate in the interface liquid-biofilm ($S_{LF,i}$) and the biofilm thickness (L_F), whereas the substrate balance in the reactor is a constraint.

For a multi-species biofilm system, the biomass density equation is used as target cell in the optimization problem, and the constraints are the substrate balances in the reactor (substrates for heterotrophs and autotrophs) and the biomass balances (for heterotrophs and autotrophs). The variables (adjustable cells) are substrate concentration of COD and N in the liquid-biofilm interface, detachment rate (b_{de}) for heterotrophs (assuming the same value for autotrophs and inerts) and biomass density for heterotrophs and autotrophs.

For the case of multiple limiting substrates, a new constraint is introduced, and the modified maximum specific substrate conversion rate ($q_{max,i,mod}$) must be declared as a new variable in the optimization problem (new adjustable cell), because the outflow substrate concentration of the non-limiting compound ($S_{non-lim,LF}$) is not known *a priori*. The constraint is

$$q_{max,i,mod} - q_{max,i} \frac{S_{LF,non-lim}}{S_{LF,non-lim} + K_{non-lim}} = 0 \qquad (3.59)$$

To compute $q_{max,i,mod}$, the concentration $S_{LF,non-lim}$ needs to be estimated, leading to an iterative process that must be solved manually. With MS Office Excel, this is solved using a guessed initial value that will be further recalculated for the different values of concentration of the non-limiting substrate found in the automatic iterative process (SOLVER tool in MS Office Excel).

3.2.4 Applications

3.2.4.1 Numerical versus analytical solutions

To analyze the results obtained with the weighted average approach, a set of different biofilm operating conditions were simulated for a heterotrophic flat biofilm in which liquid phase is considered to be completely mixed. The values obtained by the weighted average approach were compared with results obtained using the substrate volumetric conversion. For numerical integration, a high accuracy orthogonal collocation method (Finlayson 1972) was used on a grid of 14 nodes in the biofilm.

For a given set of parameters, the mass balances described by equations (3.46) and (3.54) have been solved to determine the steady state in a wide range of inflow substrate concentrations. Two typical situations have been chosen: the description of an existing biofilm reactor, or the design of a new biofilm reactor.

3.2.4.2 Describing an existing reactor system

When an existing reactor is described the reactor volume and influent are the input parameters for the calculation (i.e., in Figure 3.4 the reactor loading is a known variable), and the effluent composition resulting from the predicted biofilm conversion is calculated. To study the deviation of the weighted average and zero-order approaches with respect to the numerical solution, the calculated flux of rate-limiting substrate and outflow substrate concentration is plotted against the reactor loading of the rate-limiting substrate.

Using the weighted-average approach, the numerical results for flux and bulk concentration of substrate can be reproduced satisfactorily (Figure 3.4) for the whole range of concentration, including the substrate concentration in the range of the substrate affinity constant (K_i). The relative deviation in calculated substrate flux is even smaller than the relative deviation in the calculated bulk concentration of substrate for a fixed inflow load to the biofilm reactor. That is to say, the influent conditions (inflow concentration of substrate) are known, and the model is applied to compute both the flux and the bulk liquid concentration of substrate.

The zero-order approach can predict flux with relatively small deviations (e.g., for reactor loading about 2 $gCOD_S \cdot m^{-2} \cdot day^{-1}$ deviation is 17% while for 14 $gCOD_S \cdot m^{-2} \cdot day^{-1}$ it is 9%) when the reactor loading is fixed (and flux and bulk concentration are calculated). The relative errors in the predicted bulk concentration of substrate are particularly larger for low S_i/K_i ratios (see dashed arrows in Figure 3.4 as an example). This is logical, since a first-order approximation assumes high S_i/K_i. It should be noted that the half-order approach should indeed not be used when substrate concentrations in the bulk liquid are around or below the substrate affinity constant.

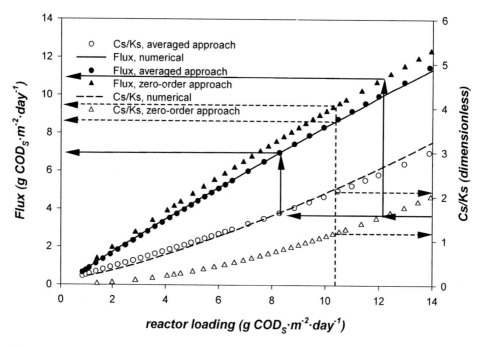

Figure 3.4. Comparison between numerical and weighted average and zero-order approaches for a steady-state biofilm in a wide range of inflow concentrations in a well-mixed tank operating in continuous mode. Arrows show differences when using the half-order approach for a fixed S_i/K_i (on the graph actually Cs/Ks) (designing a biofilm reactor, solid arrows), and for a fixed reactor loading (describing existing reactor systems, dashed arrows).

3.2.4.3 Designing a biofilm reactor

In biofilm reactor design, the influent characteristics and effluent requirements are input to the calculation of the desired reactor volume, which is calculated from the estimated required surface area and the flux of substrate towards the biofilm. A comparison of model calculations can be made when the substrate flux is estimated with the model by fixing the effluent concentration of substrate (i.e. the bulk liquid concentration of substrate). This comparison is shown in Figure 3.4 (solid arrows). It is possible to estimate the flux either by fixing the reactor loading or using as fixed the bulk concentration of substrate. Figure 3.4 illustrates that flux deviations for describing versus designing applications are relevantly different.

3.3 PSEUDO-ANALYTICAL MODELS (PA)

Closed-form analytical solutions are available only for highly simplified situations: e.g., uniform biomass density, known biofilm thickness, no external mass-transport resistance, and single-substrate limitation following first-order or zero-order kinetics. A *pseudo-analytical* (PA) model is a simple alternative when one or more of the simplifications must be eliminated to gain a realistic representation of the biofilm system. Pseudo-analytical solutions are comprised of a small set of algebraic equations that can by solved directly by hand or with a spreadsheet. The solution outputs the substrate flux (j_F) when the bulk-liquid substrate concentration (S_B) is input to it. The relative ease of using the pseudo-analytical solutions makes them amenable for routine application in process design and as a teaching tool (Grady *et al.* 1999; Rittmann and McCarty 2001). The pseudo-analytical solution is simply coupled with a reactor mass balance so that the unique combination of substrate concentration and substrate flux in the reactor is computed for a given biofilm system.

Pseudo-analytical solutions evolved by eliminating one, two, or three of the simplifications listed in the previous paragraph. Atkinson and Davies (1974) and Atkinson and How (1974) found a pseudo-analytical solution when non-linear Monod kinetics represents single-substrate limitation. Rittmann and McCarty (1981) coupled external mass-transport resistance to Monod kinetics. Rittmann and McCarty (1980) also added a mass balance on biomass so that their pseudo-analytical solution for a steady-state biofilm predicts the steady-state biofilm thickness. Sáez and Rittmann (1992) presented a more accurate pseudo-analytical solution for steady-state biofilms, and it has been incorporated into recent textbooks that cover biofilm kinetics (Grady *et al.* 1999; Rittmann and McCarty 2001).

The pseudo-analytical solution for a steady-state biofilm (Sáez and Rittmann 1992) can be applied for multi-species biofilms, a new feature necessary for BM3 (Section 4.4). Such a pseudo-analytical solution makes multi-species modeling more accessible to students, engineers, and non-specialist researchers. In addition, creating and using a multi-species pseudo-analytical model illuminates the important interactions that take place among the different types of biomass in a multi-species biofilm.

3.3.1 Features of the basic pseudo-analytical model

The basic pseudo-analytical model is for a steady-state biofilm with one microbial species and one rate-limiting substrate (Rittmann and McCarty 1980, 2001; Sáez and Rittmann 1992). This is the PA model used for BM1 (Section 4.2) and BM2 (Section 4.3). The process considered in the steady-state biofilm model are substrate utilization according to the single-Monod expression in the biofilm, transport by diffusion perpendicular to the substratum, transport by diffusion through an external diffusion layer (or boundary layer), and a steady-state mass balance on active biomass in the biofilm. The pseudo-analytical solution was attained by fitting thousands of numerical solutions of this set of equations to create the set of algebraic equations comprising the pseudo-analytical solution.

The algebraic equations that comprise the pseudo-analytical solution of the steady-state biofilm reactor are greatly simplified by defining the following dimensionless variables, denoted with an overscript (~).

$$\tilde{S}_{in} = \frac{S_{in}}{K_S} \qquad \tilde{S}_{LF} = \frac{S_{LF}}{K_S} \qquad \tilde{S}_B = \frac{S_B}{K_S} \qquad b = b_{ina} + b_{res} + b_{de}$$

$$\tilde{S}_{min} = \frac{b}{Y q_{max} - b} \qquad \tilde{K} = \frac{D}{L_L}\left[\frac{K_S}{q_{max} X_F D_F}\right]^{1/2} \qquad \tilde{R} = \frac{D}{L_L}\frac{A_F}{Q} \qquad \tilde{j}_F = \frac{j_F}{\left(K_S q_{max} X_F D_F\right)^{1/2}}$$

The pseudo-analytical solution for a steady-state biofilm is expressed as

$$\tilde{j}_F = f \, \tilde{j}_{deep} \tag{3.60}$$

where $\tilde{j}_{deep}$ is the dimensionless flux into a deep biofilm and can be calculated analytically with

$$\tilde{j}_{deep} = \left(2\left[\tilde{S}_{LF} - \ln\left(1 + \tilde{S}_{LF}\right)\right]\right)^{1/2} \tag{3.61}$$

Sáez and Rittmann (1992) found that

$$f = \tanh\left[\alpha\left(\frac{\tilde{S}_{LF}}{\tilde{S}_{min}} - 1\right)^{\beta}\right] \tag{3.62}$$

where

$$\alpha = 1.5557 - 0.4117 \tanh\left(\log_{10} \tilde{S}_{min}\right)$$

$$\beta = 0.5035 - 0.0257 \tanh\left(\log_{10} \tilde{S}_{min}\right)$$

By substituting equations (3.61) and (3.62) in equation (3.60), the dimensionless flux-substrate relationship for the biofilm side becomes

$$\tilde{j}_F = \tanh\left[\alpha\left(\frac{\tilde{S}_{LF}}{\tilde{S}_{min}} - 1\right)^{\beta}\right]\left(2\left[\tilde{S}_{LF} - \ln\left(1 + \tilde{S}_{LF}\right)\right]\right)^{1/2} \tag{3.63}$$

If the dimensionless concentration at the outer surface of the biofilm is known ($\tilde{S}_{LF}$), the dimensionless flux ($\tilde{j}_F$) is computed with equation (3.63), once the values of $\tilde{S}_{min}$, α, and β have been calculated. To know $\tilde{S}_{LF}$, one normally must couple a reactor mass balance to equation (3.63) by using the flux across the external diffusion layer as a common boundary condition. The dimensionless steady-state mass balance for a single completely mixed biofilm reactor is

$$0 = \left(\tilde{S}_{in} - \tilde{S}_B\right) - \tilde{R}\left(\tilde{S}_B - \tilde{S}_{LF}\right) \tag{3.64}$$

Using Fick's law for the liquid film, the dimensionless flux into the biofilm, $\tilde{j}_F$, can be written as

$$\tilde{j}_F = \tilde{K}\left(\tilde{S}_B - \tilde{S}_{LF}\right) \tag{3.65}$$

Substituting equation (3.65) into equation (3.64) connects the dimensionless flux ($\tilde{j}_F$) to the dimensionless substrate concentration input to the reactor ($\tilde{S}_{in}$):

$$\tilde{j}_F = \frac{\tilde{K}}{1+\tilde{R}}\left(\tilde{S}_{in} - \tilde{S}_{LF}\right) \qquad (3.66)$$

Thus, the steady-state biofilm reactor solution normally is computed by equating the fluxes of expressions (3.63) and (3.66) and then determining $\tilde{S}_{LF}$ using a root finder, such as the GOAL SEEK function within EXCEL. The computation also can be accomplished *by hand* by assuming values of $\tilde{S}_{LF}$ and iterating until the same $\tilde{j}_F$ value is obtained in (3.63) and (3.66). In addition, the solution can be graphical by intersecting the curves given by expressions (3.63) and (3.66).

Once values of the dimensionless concentrations and fluxes are computed, they can be converted to the dimensional domain to give S_B, S_{LF}, and j_F by using the definitions of the dimensionless parameters. Then, other important quantities can be determined directly from the output values of the basic pseudo-analytical solution. The total flux of oxygen for a heterotrophic biofilm is

$$j_{F,O2} = j_{F,S}\left(1 - Y_H\right) + b_{res,H}\, X_{F,H} L_{F,H} \qquad (3.67)$$

The biofilm accumulations of heterotrophs inerts are given by

$$X_{F,H} L_{F,H} = j_{F,S}\frac{Y_H}{b_H}, \qquad X_{F,I} L_{F,I} = X_{F,H} L_{F,H}\frac{b_{ina,H}}{b_{de,I}} \qquad (3.68)$$

3.3.2 Adapting the pseudo-analytical model for multiple species

As preparation to address BM3 in Section 4.4, we adapted the basic PA model to describe a multi-species biofilm has three biomass types and two rate-limiting substrates:

• The biomass types are aerobic heterotrophs (X_H), autotrophic nitrifiers (X_N), and inert biomass (X_I).

• The substrates are organic chemical oxygen demand (COD) for the heterotrophs (S_S) and NH$_3$-N for the nitrifiers (S_N). Dissolved oxygen (S_{O2}) is the electron acceptor for both forms of active biomass, but we assume that it is not rate limiting for any reactions in this work.

• Inert biomass is produced as part of decay of the active biomass.

Table 3.1 is a matrix that shows the stoichiometric relationships among the substrates and biomass types, as well as the rate expressions used. Table 3.2 defines the standard stoichiometric and kinetic parameters and lists their values and units. Table 3.2 also lists the standard system parameters. The system is one completely mixed biofilm reactor; suspended biomass does not contribute to any reactions.

3.3.3 The multi-species models

We employed four levels of multi-species modeling using the steady-state-biofilm model (Sáez and Rittmann 1992) for each. The models, designated PAa through PAd, involve increasing degrees of interaction among the biomass types. Table 3.3 summarizes the distinguishing features among the four models. One of the outcomes of this work is to identify when and how the added interactions significantly affect model results. Knowing when added interactions have a significant effect on results is important, because the added interactions force the spreadsheets to become more sophisticated.

Table 3.1 Matrix of stoichiometry and rate expressions used for multi-species PA modeling

Process	Biomass Types			Substrates			Kinetic Expressions
name	X_H	X_N	X_I	S_S	S_N	S_{O2}	
Heterotroph metabolism	1			$\dfrac{-1}{Y_H}$		$\dfrac{-(1-Y_H)}{Y_H}$	$\mu_{max,H}\dfrac{S_S}{K_S+S_S}X_H$
Heterotroph inactivation	-1		1				$b_{ina,H}\,X_H$
Heterotroph respiration	-1					-1	$b_{res,H}\,X_H$
Autotroph metabolism		1			$\dfrac{-1}{Y_N}$	$\dfrac{-(4.57-Y_N)}{Y_N}$	$\mu_{max,N}\dfrac{S_N}{K_N+S_N}X_N$
Autotroph inactivation		-1	1				$b_{ina,N}\,X_N$
Autotroph respiration		-1				-1	$b_{res,N}\,X_N$

Table 3.2. Standard kinetic, stoichiometric, and system parameters for applying the multi-species PA models

Description	Symbol	Units	Value	Symbol	Units	Value
Stoichiometric and Kinetic Parameters						
Maximum growth rates	$\mu_{max,H}$	1/d	6.0	$\mu_{max,N}$	1/d	1.0
Half-maximum-rate concentrations	K_s	g_{CODS}/m^3	4.0	K_N	g_N/m^3	1.0
True yields	Y_H	g_{CODX}/g_{CODS}	0.63	Y_N	g_{CODX}/g_N	0.24
Inactivation rate coefficients	$b_{ina,H}$	1/d	0.08	$b_{ina,N}$	1/d	0.03
Respiration rate coefficients	$b_{res,H}$	1/d	0.32	$b_{res,N}$	1/d	0.12
Detachment rate coefficients	$b_{de,H}$	1/d	0.4	$b_{de,N}$	1/d	0.4
Diffusion coefficients in water	D_S	m^2/d	1.0×10^{-4}	D_N	m^2/d	1.7×10^{-4}
				D_{O2}	m^2/d	2.0×10^{-4}
Diffusion coefficients in biofilm	$D_{F,S}$	m^2/d	1.0×10^{-4}	$D_{F,N}$	m^2/d	1.7×10^{-4}
				$D_{F,O2}$	m^2/d	2.0×10^{-4}
System Parameters						
Influent substrate concentration	$S_{in,S}$	g_{CODS}/m^3	30	$S_{in,N}$	gN/m^3	6
Flow rate and Reactor volume	Q	m^3/d	0.02	V	m^3	0.00125
Biofilm surface area and Diffusion-layer thickness	A_F	m^2	0.1	L_L	m	10^{-8}
Maximum biofilm density				X_F	g_{CODX}/m^3	10^4

Biomass is expressed as solid COD, or subscript CODX. Organic substrate is expressed as soluble COD, or subscript CODS. Ammonia N is expressed as N.

Table 3.3. Distinguishing features among the four PA models for multi-species biofilm

Model	Distinguishing Features
PAa	No interactions between heterotrophs and nitrifiers. Fluxes and concentrations of COD and NH3-N are computed independently.
PAb	The only interaction is N consumption for heterotroph net synthesis. The nitrifying biomass produced is computed from net biomass production and included in the NH3-N mass balance.
PAc	Interactions are N consumption for heterotroph net synthesis and space competition. Nitrifiers and inerts exist in layers behind the heterotrophs. NH3-N has increased mass-transport resistance through the heterotroph layer. Nitrifiers and inerts experience reduced detachment rates.
PAd	Interactions are N consumption for heterotroph net synthesis and space competition. Densities of heterotrophs, nitrifiers, and inerts sum to the maximum biomass density according to their ability to compete.

Model PAa -- No Interactions. The first model treats the heterotrophs and nitrifiers completely independently. This model is essentially the same approach as taken by Rittmann and Stilwell (2002) for their Integrated Biofiltration Model, although they also included soluble microbial products from heterotrophs and nitrifiers, an issue not part of the evaluation here or with BM3. The pseudo-analytical solution is solved simultaneously and independently with a steady-state reactor mass balance for each of COD and NH$_3$-N. The two solutions each give outputs of bulk-liquid substrate concentration (S_S or S_N), substrate flux ($j_{F,S}$ or $j_{F,N}$), and biofilm accumulation per unit area ($X_F L_{F,H}$ or $X_F L_{F,N}$). The accumulation of inert biomass is then computed from

$$X_F L_{F,I} = (b_{ina,H} X_F L_{F,H} + b_{ina,N} X_F L_{F,N}) / b_{de} \qquad (3.69)$$

In addition, the oxygen flux ($j_{F,O2}$) is computed at the end from

$$j_{F,O2} = j_{F,S}(1 - Y_H) + b_{res,H} X_F L_{F,H} + j_{F,N}(4.57 g_{O_2}/g_N - Y_N) + b_{res,N} X_F L_{F,N} \qquad (3.70)$$

Model PAb -- Competition for N. Model PAb introduces the first level of interaction, the competition between heterotrophs and nitrifiers for NH$_3$-N. The nitrifiers use NH$_3$-N as their electron acceptor and as the N source for synthesis; the heterotrophs also utilize NH$_3$-N for synthesis. When the production of heterotrophic biomass is large, its net synthesis can be a large sink that competes with utilization for nitrification. To include competition for N, we add for Model PAb the flux of NH$_3$-N for heterotroph net synthesis:

$$j_{F,N/H} = (0.0872 g_N/g_{CODX})(Y_H j_{F,S} - X_F L_{F,H} b_{res}) \qquad (3.71)$$

This flux is then added to the steady-state reactor mass balance for NH$_3$-N:

$$0 = Q S_{in,N} - Q S_{B,N} - A_F (j_{F,N} + j_{F,N/H}) \qquad (3.72)$$

Adding the second flux sink term to the mass balance represents competition for NH$_3$-N and causes $j_{N,F}$ to decrease to compensate.

Model PAc -- Layering and Protection. Model PAc adds a different type of interaction, the competition for space. In Model PAc, the three different types of biomass exist in parallel layers. Nearest the outer surface are the heterotrophs. Behind the heterotrophs are the nitrifiers. The inert biomass lies behind the nitrifiers and next to the substratum. This layering approach is based on prior results of multi-species modeling (e.g., Wanner and

Gujer 1986; Rittmann and Manem 1992; Furumai and Rittmann 1994; Rittmann *et al.* 2001) and experiments (Rittmann and Manem 1992; Watanabe *et al.* 1995; Rittmann *et al.* 1992) that show that the faster growing heterotrophs tend to dominate the outer part of a biofilm, while nitrifiers and inerts exist closer to the substratum. Using parallel layers is a simple means to represent the tendency of heterotrophs to be mostly near the outer surface.

Implementing Model PAc requires two modifications to the set up of Model PAb. First, the heterotrophic layer becomes an extra diffusion layer for transport of NH_3-N to the nitrifier layer. Modifying the boundary condition for the nitrifying layer represents the extra resistance:

$$j_{N,F} = \frac{D_N D_{N,F} (S_N - S_{N,in})}{D_{N,F} L_L + D_N L_{F,H}}$$

(3.73)

in which $L_{F,H}$ = the thickness of the (outer) heterotroph layer, and $S_{m,N}$ = the concentration of NH_3-N at the outer surface of the nitrifier layer. Second, the nitrifier and inert biomasses are protected from detachment loss, since they are away from the outer surface. We represent this form of protection by

$$b_{de,N} = f_N \, b_{de,H} \quad \text{and} \quad b_{de,I} = f_I \, b_{de,H}$$

(3.74)

in which f_N and f_I are fractions less than one that represent the degree to which the nitrifiers and inerts have reduced detachment rates and the subscripts H, N, or I indicate a detachment rate unique to that biomass type. For the results show here, we set $f_N = f_I = 0.1$. The solution requires that the composite detachment rate for all types of biomass be equal to the overall detachment rate of b_{det}. Mathematically,

$$b_{de} = \frac{b_{de,H} X_F L_{F,H} + b_{de,N} X_F L_{F,N} + b_{de,I} X_F L_{F,I}}{X_F L_{F,H} + X_F L_{F,N} + X_F L_{F,I}}$$

(3.75)

Solving Model PAc this way requires iteration on $b_{de,H}$ until the b_{de} value computed by Equation 7 equals the desired overall detachment rate b_{de}, such as 0.4/d for the standard condition (Table 3.2).

Model PAd -- Competitive Space Distribution. Model PAd incorporates the effects of space competition in a different way than does Model 3. In Model 4, the total biofilm density is comprised of the sum of the densities of each biomass type:

$$X_F = X_{F,H} + X_{F,N} + X_{F,I}$$

(3.76)

This distribution of densities is constant throughout the biofilm. To compute the densities, we begin with the results of Model PAb, which gives $X_F L_F$ values for each biomass type. We sum the three values and compute the fractions of that total attributed to each biomass type. For the heterotrophs and nitrifiers, we use the fractions to reduce their X_F values from the maximum value. For example, if the heterotrophs comprise 50% of the total $X_F L_F$ from the results of Model PAb, we make $X_{F,H} = 0.5 X_F = 5000$ g_{CODX}/m^3. With the reduced X_F values for heterotrophs and nitrifiers, we compute new values for S_S, S_N, $j_{F,S}$, $j_{F,N}$, and the three $X_F L_F$ values. If the newly computed values of $X_F L_F$ are the same as before, the computations end, and we have the Model PAd output. If the $X_F L_F$ values differ from the previous set of computations, we again compute fractions, reduce the X_F values, and run the model. The results normally converge within three iterations.

When running Model PAd for the work reported here, we did not adjust the b_{det} values to reflect any protection of the heterotrophs and nitrifiers. Although this merging of Models 3 and 4 would be simple to implement, it creates a contradiction. Having the nitrifiers and inerts protected by being behind the heterotrophs is inconsistent with dividing X_F uniformly among the three biomass types. Thus, Models PAc and PAd provide limiting representations of space competition: segregation and protection for Model PAc versus unmitigated competition for Model PAd. In this sense, the results of the two models may define the range of results defined by space competition.

3.3.4 Multi-species applications

To illustrate how the four multi-species PA models work and the insight that can be gained by using, them, we present at series of example applications. All are for one completely mixed biofilm reactor with the conditions of Table 3.2. The standard condition corresponds exactly to the parameters listed in Table 3.2 and is designed to allow a significant accumulation of all three biomass types. A series of special conditions change one parameter to test different scenarios of practical and theoretical interest.

3.3.4.1 Standard condition

Table 3.4 summarizes the key outputs of the four models for the standard condition, which is shown in Table 3.2 and represents a typical COD:N ratio (i.e., 5 g_{CODS}/gN) found in domestic wastewater. The results in Table 3.4 show three main trends for how the different types of interactions affect substrate removals and the biomass types.

- The removal of COD is affected only by direct competition for space in Model PAd. The reduction of $X_{F,H}$ slows the substrate-utilization kinetics per unit volume of biofilm, which affects S_S and $j_{F,S}$.
- The removal of NH$_3$-N is affected significantly in this case by direct space competition lowering $X_{F,N}$ (Model PAd).
- Protection by layering significantly increases the accumulation of nitrifiers and inerts at the expense of heterotrophs. However, substrate fluxes and removals are hardly affected, since the nitrifiers are layered behind the heterotrophs.

Table 3.4. Summary of key output parameters for the standard condition for the multi-species PA models

Model	S_S g_{CODS}/m^3	S_N gN/m^3	$j_{F,S}$ g_{CODS}/m^{2}d	$j_{F,N}$ g_N/m^{2}d	$X_FL_{F,H}$ g_{CODX}/m^2	$X_FL_{F,N}$ g_{CODX}/m^2	$X_FL_{F,I}$ g_{CODX}/m^2	L_F μm
PAa	4.36	1.43	5.13	0.91	4.06	0.40	0.84	530
PAb	4.36	1.41	5.13	0.92	4.06	0.33	0.84	522
PAc	4.37	1.51	5.13	0.90	**2.65**	**0.72**	**2.85**	**622**
PAd	**4.93**	**2.00**	**5.01**	**0.80**	3.96	0.28	0.81	505

For each column, the **boldface** entries are significantly different and are discussed in the text.

3.3.4.2 High influent N:COD

Table 3.5 summarizes the key outputs for a condition of a high influent N:COD ratio. The influent NH$_3$-N concentration is increased to 30 mg/L, with the influent COD concentration still at 30 mg/L. One novel trend is evident when the potential to grow nitrifiers was much greater. Space competition (Models PAc and PAd) adversely affects the removal of NH$_3$-N. For the layering of Model PAc, the negative impact comes about from the added mass-transport resistance through the heterotrophic layer, since the total accumulation of nitrifiers was increased by protection. Compared to the results of the standard condition (Table 3.4), the impact of added resistance is much more significant for the high N:COD ratio, because the potential NH$_3$-N flux is about five times greater. On the other hand, direct space competition in Model PAd causes a reduction in the accumulation of nitrifiers, and this causes reduced removal of NH$_3$-N.

Table 3.5. Summary of key output parameters for the condition of a high influent N:COD ratio using the multi-species PA models

Model	S_S g_{CODS}/m^3	S_N gN/m^3	$j_{F,S}$ g_{CODS}/m^2d	$j_{F,N}$ gN/m^2d	$X_F L_{F,H}$ g_{CODX}/m^2	$X_F L_{F,N}$ g_{CODX}/m^2	$X_F L_{F,I}$ g_{CODX}/m^2	L_F μm
PAa	4.36	4.29	5.13	5.14	4.06	2.25	0.98	728
PAb	4.36	3.78	5.13	5.24	4.06	2.22	0.98	725
PAc	4.38	**20.75**	5.12	**1.85**	**2.43**	1.62	**2.60**	665
PAd	5.31	**17.51**	4.94	**2.50**	3.90	**1.02**	0.86	**578**

For each column, the **boldface** entries are significantly different and are discussed in the text.

3.3.4.3 Low influent N:COD

Table 3.6 contains the output results for a condition in which the influent N:COD ratio is very low: 1.5 g_N/m^3:$30g_{CODS}/m^3$. In this situation, net synthesis of heterotrophic biomass consumes most of the NH_3-N, eliminating nitrification and nitrifiers in Models PAb and PAd. Comparing the results of Model PAa to the other results shows that $j_{F,NH}$ is a major sink for NH_3-N in this case. With a low N:COD ratio, the impact of protection by layering is amplified, and Model PAc gives considerably greater accumulation of inerts and some nitrifiers.

Table 3.6. Summary of key output parameters for the condition of a low N:COD ratio using the multi-species PA models

Model	S_S g_{CODS}/m^3	S_N gN/m^3	$j_{F,S}$ g_{CODS}/m^2d	$j_{F,N}$ gN/m^2d	$X_F L_{F,H}$ g_{CODX}/m^2	$X_F L_{F,N}$ g_{CODX}/m^2	$X_F L_{F,I}$ g_{CODX}/m^2	L_F μm
PAa	4.36	**1.36**	5.13	*0.03*	4.06	0.01	0.81	488
PAb	4.36	0.65	5.13	0.17	4.06	0.00	0.81	487
PAc	4.37	**0.35**	5.13	0.23	**2.82**	0.03	**3.03**	**588**
PAd	4.78	0.66	5.04	0.17	3.99	0.00	0.80	479

For each column, the **boldface** entries are significantly different and are discussed in the text.

3.3.4.4 High detachment rate

The results of the final condition investigated are given in Table 3.7. Here, the global detachment rate (b_{de}) is increased to 2.15/d. For Model PAc, $b_{de,H}$ was 4.5/d to achieve the composite value of $b_{de} = 2.15$/d. The high-detachment condition yields three novel trends.
- The nitrifiers are washed out of the biofilm unless layering (Model PAc) protects them.
- When nitrifiers are washed out, the only sink for NH_3-N is heterotroph net synthesis. Hence, the S_N and $j_{F,N}$ values are different for Model PAa, which has no N synthesis.
- Ironically, protection by layering in this case causes the biofilm thickness to shrink. The reason is that the faster growing heterotrophs are exposed to a very large detachment rate ($b_{de,H} = 4.5$/d), which greatly reduces their accumulation.

Table 3.7. Summary of key output parameters for the condition of a high detachment rate using the multi-species PA models

Model	S_S g_{CODS}/m^3	S_N gN/m^3	$j_{F,S}$ g_{CODS}/m^2d	$j_{F,N}$ gN/m^2d	$X_F L_{F,H}$ g_{CODX}/m^2	$X_F L_{F,N}$ g_{CODX}/m^2	$X_F L_{F,I}$ g_{CODX}/m^2	L_F μm
PAa	5.07	**6.00**	4.99	0.00	1.23	0.00	0.05	128
PAb	5.07	4.80	4.99	0.24	1.23	0.00	0.05	128
PAc	**19.93**	1.85	2.01	**0.83**	0.26	**0.29**	0.07	**62**
PAd	5.13	4.80	4.97	0.24	1.23	0.00	0.05	128

For each column, the **boldface** entries are significantly different and are discussed in the text.

3.3.4.5 Oxygen Flux

Table 3.8 summarizes the total oxygen flux, as computed by Equation (3.70), for all conditions and models. Three significant trends are evident.

- For the standard condition, the highest oxygen flux is with Model PAa, because no NH_3-N is diverted to synthesis of heterotrophic biomass.
- The model makes the greatest difference for the condition of a high influent N:COD ratio, where the potential oxygen consumption by nitrification is 4.57 times greater than for COD oxidation. The extra mass-transport resistance to move NH_3-N through the layer of heterotrophs (Model PAc) greatly reduces $j_{F,N}$ (Table 3.8), which translates to a much reduced $j_{F,O2}$. The significantly lowered accumulation of nitrifiers for Model PAd (Table 3.5) also affects $j_{F,N}$ and $j_{F,O2}$.
- On the other hand, the protection afforded by layering in Model PAc allows some nitrification to take place for the high-detachment condition (Table 3.8), and this makes the oxygen flux largest with Model 3.

Table 3.8. Oxygen fluxes ($j_{F,O2}$ in gO_2/m^2d) for all conditions and multi-species PA models

Condition	Model PAa	Model PAb	ModelPAm 3	Model PAm 4
Standard	**7.18**	6.46	5.81	5.89
High N:COD	25.71	25.42	**9.94**	**13.29**
Low N:COD	3.31	3.18	2.91	3.13
High Detachment	2.23	2.24	**4.00**	2.23

For each row, the **boldface** entries are significantly different and are discussed in the text.

3.3.4.6 Interfacial Concentrations and Biofilm Deepness

From the substrate flux and bulk-liquid concentrations, we can compute substrate concentrations at the outer surface of the biofilm, $S_{LF,N}$ and $S_{LF,S}$, from a rearranged version of Equation (3.73):

$$S_{LF,N} = S_{B,N} - j_{F,N} \frac{L_L}{D_N} \tag{3.77a}$$

$$S_{LF,S} = S_{B,S} - j_{F,S} \frac{L_L}{D_S} \tag{3.77b}$$

For Model PAc, the nitrogen inter-layer concentration $S_{m,N}$ can be computed from:

$$S_{m,N} = S_{LF,N} - j_{F,N} \frac{L_{F,H}}{D_{F,N}} \tag{3.78}$$

The substrate concentration at the base of the biofilm for each substrate can then be computed from a version of equation (3.63) (Rittmann and McCarty 2001), which is for COD:

$$j_{F,S} = \left[2 q_{max,H} X_{F,H} D_{F,S} \left(S_{LF,S} - S_{0,S} + K_S \ln \frac{K_S + S_{0,S}}{K_S + S_{LF,S}} \right) \right]^{1/2} \tag{3.79}$$

Once the pseudo-analytical solution has been solved, the only unknown value in Equation (3.79) is $S_{0,S}$, which is the concentration of substrate (COD in this case, but also for NH_3-N) at the substratum, which also is at the inside of the heterotroph or nitrifier layer in Model PAc. Table 3.9 summarizes the interfacial concentrations for the standard case. For COD, the values of $S_{0,S}$ are much less than for $S_{LF,S}$, and this indicates that the biofilm is relatively *deep* for the standard case. On the other hand, the values of $S_{0,N}$ are only slightly smaller than

$S_{LF,N}$, and the biofilms are *shallow* for NH$_3$-H. The biofilm is less deep for COD with Model PAc, because $X_F L_{F,H}$ is reduced in Model PAc due to the higher detachment rate for heterotrophs, which are at the outer surface of the biofilm (Table 3.3). $S_{LF,N}$ and $S_{0,N}$ are substantially lower for Model PAc, since NH$_3$-N has to diffuse through the heterotrophic layer before reaching the nitrifier layer. Model PAd gives higher surface concentrations, but lower base concentrations, compared to Models PAa and PAb. The reason is that the thickness of the heterotroph and nitrifiers layers is greater, since they compete for the same space, making up a composite X_F.

Table 3.9. Interfacial concentrations for the standard case using the multi-species PA models

Concentration	Model PAa	Model PAb	Model PAc	Model PAd
$S_{LF,S}$	4.36	4.36	4.37	4.93
$S_{0,S}$	0.066	0.066	0.27	0.051
$S_{LF,N}$ [1]	1.43	1.41	0.43	2.00
$S_{0,N}$	1.33	1.34	0.29	1.11

[1] $S_{m,N}$ for Model PAc.

3.3.5 Summary for multi-species PA models

It is possible to adapt the PA model for a steady-state biofilm to describe multi-species biofilms. The major advantage of this approach is that the multi-species models can be solved with a simple spreadsheet. This opens multi-species modeling to students, practicing engineers, and non-specialist researchers.

Four PAm models have increasing degrees of interaction between aerobic heterotrophs and nitrifiers. Model PAa has no interactions, which means that the substrate concentration, substrate flux, and biomass accumulation are computed separately for heterotrophs and nitrifiers. Model PAb adds consumption of NH$_3$-N as an N source for heterotrophs. Models PAc and PAd include space competition among the heterotrophs, nitrifiers, and inert biomass. Model PAc assumes that the biomass types exist in separate layers, with heterotrophs at the outer surface of the biofilm and inert biomass at the substratum. This layering adds mass transport resistance to supply the NH$_3$-N to the nitrifier layer, but it also protects the nitrifiers and inerts from detachment. Model PAd assumes that the three biomass types directly compete for their share of the maximum biomass density in the biofilm.

Modeling results for four conditions identify the effects of the interactions present in each model and when they have a significant impact. For example, not including N uptake for synthesis has its greatest impact when the nitrifiers comprise a small fraction of the biofilm; a high detachment rate or a low influent N:COD ratio created the condition making synthesis an important N sink compared to nitrification. Protection by layering has its greatest positive impact on nitrification when the detachment rate is high and nitrifiers are washed out without this type of protection. On the other hand, protection by layering significantly slows the rate of NH$_3$-N removal when the flux of NH$_3$-N has the potential to be high due to an elevated influent NH$_3$-N concentration. Finally, direct competition for space in all parts of the biofilm (Model PAd) tends to slow the removal rates for COD and NH$_3$-N, because both biomass types are "diluted" by the presence of the other.

3.4   NUMERICAL ONE-DIMENSIONAL DYNAMIC MODEL (N1)

The model referred to as N1 is a multi-species and multi-substrate model that represents the biofilm in one dimension (1d) perpendicular to the substratum. Its complexity lies between the simpler (pseudo)analytical models and the numerically demanding multi-dimensional models.

The N1 model equations must be solved numerically, but even complex simulations can be performed on a PC within minutes. The most significant feature of the N1 model is its flexibility with regard to the number of dissolved and particulate components, the microbial kinetics, and to a certain extent also the physical and geometrical properties of the biofilm. This flexibility is offered by the simulation program AQUASIM (Reichert 1998a, 1998b), in which the N1 model often is implemented and by which very efficiently alternative versions of a model can be tested and experimental data can be evaluated. The N1 model can be used as a tool in research, as well as for the design and simulation of biofilm reactors.

3.4.1 Features

In the N1 model, different types of variables are treated differently: the attached particulate components, which form the biofilm solid matrix, versus the suspended particulate and the dissolved components, which exist in the biofilm liquid phase and in the bulk liquid. Examples of particulate components are active microbial species, organic and inorganic particles, and EPS. Examples of dissolved components are organic and inorganic substrates, metabolites, products, and the hydrogen ion.

The output produced by the model includes
- spatial profiles of any number of particulate components in the biofilm
- accumulation and the loss from the system of the mass of the particulate components
- spatial profiles of any number of dissolved components in the biofilm
- removal rates and effluent concentrations of the dissolved components
- biofilm thickness as a function of the production and decay of particulate material in the biofilm and of attachment and detachment of cells and particles at the biofilm surface and in the biofilm interior

For all these quantities, the development in time, as well as steady state solutions, can be calculated.

The processes considered in the model include
- many transformation processes (Section 2.4.1)
- advection and diffusion of attached particulate components in the biofilm solid matrix
- attachment and detachment of particulate components at the biofilm surface and in the biofilm interior
- diffusion of suspended particulate and dissolved components in the biofilm liquid phase and in the liquid boundary layer at the biofilm surface
- complete mixing of suspended particulate and dissolved components in the bulk liquid

The data required by the model for all components include the diffusion coefficients in pure water and in the biofilm, rate laws, kinetic parameters, and stoichiometric coefficients for the transformation processes (Section 2.4.1). Furthermore, the volume fraction of the biofilm liquid phase and the density of the particulate components, as well as expressions for their attachment and detachment velocities, must be given. The geometric and hydraulic data needed are the biofilm surface area, the bulk liquid volume, the thickness of the liquid boundary layer, and the inflow rate.

To start a simulation, the influent and bulk liquid concentrations of all components, the initial spatial distribution of the particulate components in the biofilm, and the initial biofilm thickness must be known. The spatial profiles of the dissolved components in the biofilm develop so fast that the choice of their initial values usually is insignificant. In situations, in which only a few dispersed cells are already attached to the substratum, it is no problem to start the simulation with a very small mass of particulate components, i.e., with an initial biofilm thickness of a fraction of a micrometer.

Special characteristics of the N1 model include physical and geometrical biofilm properties that may change with time and space. A variable volume fraction of the biofilm liquid phase can be used to reproduce the observation that the biofilm density changes with time and space (Zhang and Bishop 1994b). Variable diffusion coefficients can be used to reproduce the observation that transport of dissolved and suspended particulate components in the biofilm can be enhanced due to advection and turbulent diffusion reaching into the biofilm (Fan *et al.* 1990; Fu *et al.* 1994; Zhang and Bishop 1994c). The equations of the N1 model are formulated in such a way that they readily can be adjusted from a flat to a spherical or cylindrical substratum.

Applications of the N1 model are possible in research, as well as in engineering. The model can be used to simulate and predict steady state solutions and the development in time of multi-species and multi-substrate biofilms. In research, the model is a tool to express hypotheses about biofilm structure and function in a form in which they can be tested experimentally and to evaluate experimental data, e.g., to determine numerical values of kinetic parameters and of physical properties of a biofilm. In engineering applications of biofilm reactors, the mass of the microbial species and the substrate removal rates and effluent concentrations can be calculated and the effects of microbial competition for space and common substrates on the reactor performance can be analyzed.

Limitations inherent to the N1 model are that gradients of variables and parameters in the biofilm are only in the direction perpendicular to the substratum and that all quantities represent averages over planes in parallel to the substratum. If the conditions in the bulk liquid are the same over the whole substratum area, biofilm properties may be about the same all over the substratum. If the substratum area is large, i.e., the substratum length and width are orders of magnitude greater than the biofilm thickness, and regular, the averages used in the model represent reasonable values. Furthermore, if the gradients in the bulk liquid in directions parallel to the substratum are small, the gradients in the biofilm in these directions probably are small compared to the gradients perpendicular to the substratum. However, on a local basis, i.e., for dimensions in the order of 100 micrometers, this is not necessarily true: If two microbial species form two adjacent clusters and if one species utilizes the product of the other, strong local spatial gradients of the product concentration should develop in the direction parallel to the substratum. The phenomenon of microbial species developing in separate clusters is termed segregation. The significance of segregation in biofilms has been investigated by Gujer (1987). If segregation is relevant, multi-dimensional biofilm models, as described in Section 3.6, must be applied.

Another limitation is due to the simplified modeling of the bulk liquid as a completely mixed compartment. By this simplification, the exact description of the flow field in the bulk liquid is excluded. However, approximate modeling of plug flow is still possible when the system is divided into a series of segments for which the biofilm can be calculated independently.

The core of the N1 model consists of a system of stiff, non-linear partial differential equations. Because of the stiffness of the equation system, integration methods tailored for stiff systems must be used, or the dissolved and the particulate components have to be treated differently (Section 2.6.3). Furthermore, biofilm growth, i.e., the displacement of the interface between biofilm and bulk liquid, creates a so-called moving boundary problem (Kissel *et al.* 1984; Wanner and Gujer 1986). Simulations with the N1 model often are performed using the software package AQUASIM (Reichert 1998a, 1998b), since it can handle stiff systems and the moving boundary. The N1 model implemented in AQUASIM is described in Section 3.4.3.

3.4.2 Definitions and equations

The equations of the N1 model are described in full detail in two papers (Wanner and Reichert 1996; Reichert and Wanner 1997). Those equations are rather complex because they consider many processes in addition to those mentioned in Section 3.4.1. An example of such a process is the displacement of water, along with the dissolved components contained in it, from the biofilm as result of the penetration of a particle into the biofilm. However, these additional processes only serve to make the mass balance equations rigorous, and they are not usually of practical significance. Therefore, in this section, only the most important definitions and equations of the model are presented, and for the sake of simplicity only those processes are included which are of practical significance.

Biofilms are multiphase systems. Consequently, in the N1 model three different phases are distinguished. The solid attached phase is made up by the particulate components, which form the biofilm solid matrix. They are defined as

$$X_{M,i} = \rho_{s,i}\, \varepsilon_{s,i} \qquad (3.80)$$

where $X_{M,i}$ is the concentration of the attached component i ($M_X L^{-3}$), $\rho_{s,i}$ is its density ($M_X L^{-3}$), defined as the mass divided by the volume of the cell or particle, and $\varepsilon_{s,i}$ is its volume fraction (-), defined as volume of the component per unit biofilm volume. The volume fraction of the biofilm solid matrix is $\Sigma \varepsilon_{s,i}$, and the porosity or biofilm pore volume fraction θ (-) is

$$\theta = 1 - \sum_{i=1}^{n_X} \frac{X_{M,i}}{\rho_{s,i}} \qquad (3.81)$$

where n_X is the number of particulate components considered in the model. The pore volume is formed by two phases: the phase of the suspended particulate components with concentrations $X_{P,i}$ ($M_X L^{-3}$) and the biofilm liquid phase, with the liquid phase volume fraction:

$$\varepsilon_{l,F} = \theta - \sum_{i=1}^{n_X} \frac{X_{P,i}}{\rho_{s,i}} \qquad (3.82)$$

One-dimensional mass balance equations for attached particulate, suspended particulate, and dissolved components can be derived from the general mass balance (equation (3.3)) if gradients are considered only in the direction of z, perpendicular to the substratum. For a dissolved substrate i, the specific mass flux in the biofilm, $j_{F,i}$ ($M_S L^{-2} T^{-1}$), can be modeled by Fick's first law of diffusion as

$$j_{F,i} = -D_{F,i} \frac{dS_{F,i}}{dz} \qquad (3.83)$$

where $D_{F,i}$ is the molecular diffusivity in the biofilm ($L^2 T^{-1}$) and $S_{F,i}$ is the concentration of the dissolved substrate ($M_S L^{-3}$). Substituting this equation into the equation (3.3) yields a mass balance equation that describes the development in time and the spatial profile in the biofilm of the concentration $S_{F,i}$ of a dissolved substrate as

$$\frac{\partial S_{F,i}}{\partial t} = D_{F,i} \frac{\partial^2 S_{F,i}}{\partial z^2} + r_{F,i} \qquad (3.84)$$

where $r_{F,i}$ is the substrate net production rate ($M_S L^{-3}$). Equation (3.84) has the boundary conditions

$$\frac{\partial S_{F,i}}{\partial z} = 0 \qquad (3.85)$$

at the substratum ($z=0$) and

$$S_{F,i} = S_{LF,i} \qquad (3.86)$$

at the biofilm – bulk liquid interface ($z=L_F$), where L_F is the biofilm thickness (L) and $S_{LF,i}$ is the substrate concentration at the biofilm surface (M_SL^{-3}). All terms in equation (3.84) are given as substrate mass per unit biofilm volume. Note that in the equations in the two papers mentioned above (Wanner and Reichert 1996; Reichert and Wanner 1997), the substrate concentration $C_{F,i}$ (ML^{-3}) is used, which is defined as substrate mass per unit biofilm liquid phase and which relates to $S_{F,i}$ by

$$S_{F,i} = \varepsilon_{l,F}C_{F,i} \qquad (3.87)$$

Equations (3.84) to (3.86) also hold for suspended particulate components, i.e., for cells or particles which are suspended in the biofilm pore volume and have the concentration $X_{P,i}$ (M_XL^{-3}).

For attached particulate components, which form the biofilm solid matrix, transport is assumed to be the result of microbial growth and decay in the biofilm. Growing or shrinking cells lead to a volume expansion or contraction of the biofilm solid matrix, respectively, and to a displacement of neighboring cells (Wanner 1989). This displacement can be interpreted as advective transport and is formally described as a specific mass flux $j_{M,i}$ ($M_XL^{-2}T^{-1}$) by

$$j_{M,i} = u_F X_{M,i} \qquad (3.88)$$

where u_F is the distance by which the cells are displaced per unit time (LT^{-1}). The displacement velocity u_F of a cell at the location z is equal to the added net specific mass production of all microbial species of the biofilm matrix between the substratum and this location:

$$u_F(z) = \frac{1}{1-\theta}\int_0^z \sum_{i=1}^{n_X}\frac{r_{M,i}}{\rho_{si}}dz' \qquad (3.89)$$

Based on equations (3.88) and (3.89), a mass balance analogous to equation (3.84) can be derived and used to describe the development in time and the spatial profile in the biofilm of the attached particulate component i:

$$\frac{\partial X_{M,i}}{\partial t} = -\frac{\partial(u_F X_{M,i})}{\partial z} + r_{M,i} \qquad (3.90)$$

The boundary condition needed to solve equation (3.90) is a no-flux condition at the substratum,

$$j_{M,i} = 0 \qquad (3.91)$$

for $z=0$. Equation (3.90) is used to calculate the relative abundance, spatial distribution, and development in time of microbial species and particles in the biofilm.

The development of the biofilm thickness in time is the result of the net production of biomass in the biofilm, as described by equation (3.89), of the attachment at the biofilm surface of microbial cells and particles suspended in the bulk liquid, and of the detachment of microbial cells and particles from the biofilm surface to the bulk liquid. It is modeled as

$$\frac{dL_F}{dt} = u_F(z = L_F) - u_{de} + u_{at} \qquad (3.92)$$

where u_{de} and u_{at} are the global detachment velocity and the attachment velocity, respectively (LT^{-1}). "Global" means that all species are detached at the same rate and that the detached mass of a microbial species is proportional to its concentration at the biofilm surface. The velocity u_{de} yields a phenomenological description of the decrease of the biofilm thickness per unit time as result of the detachment process, i.e., erosion or sloughing.

$$u_{de} = u_{de}(L_F, u_F(L_F), t, \tau_{LF},...) \qquad (3.93)$$

In addition to global modeling of detachment by u_{de}, it is also possible in the N1 model to have individual detachment of the various particulate components of the biofilm solid matrix (Wanner and Reichert 1996).

Attachment refers to the adsorption of microbial cells suspended in the bulk liquid to the biofilm surface and is modeled by an attachment velocity as

$$u_{at} = \frac{1}{1-\theta(L_F)} \sum_{i=1}^{n_X} \frac{k_{at,i} X_{L,i}}{\rho_{s,i}} \qquad (3.94)$$

where $k_{at,i}$ is the attachment rate coefficient (LT^{-1}) and $X_{L,i}$ is the concentration at the bulk liquid side of the biofilm surface of the suspended particulate component i ($M_X L^{-3}$).

The environment of the biofilm is modeled as a completely mixed volume of water, termed bulk liquid. Conversion processes in the bulk liquid can be equally important as those that take place in the biofilm. For each dissolved and particulate component considered, an additional mass balance equation of the form

$$\frac{d(V_B C_{B,i})}{dt} = Q(C_{in,i} - C_{B,i}) + A_F j_{F,i} + V_B r_{B,i} \qquad (3.95)$$

is needed, where $C_{in,i}$ and $C_{B,i}$ are the influent and bulk liquid concentrations, respectively, of the dissolved or suspended particulate component i (ML^{-3}), V_B is the bulk liquid volume (L^3), Q is the rate of flow through the bulk liquid ($L^3 T^{-1}$), A_F is the biofilm surface area (L^2), $j_{F,i}$ is the mass flux across the biofilm surface ($ML^{-2}T^{-1}$), and $r_{B,i}$ is the production rate ($ML^{-3}T^{-1}$) of the component.

The exchange of dissolved and particulate components between the biofilm and the bulk liquid is usually affected by the existence of a liquid boundary layer, which creates a mass transfer resistance outside the biofilm surface and in which mass transport is due to molecular diffusion only. This mass transport can be modeled as

$$j_{F,i} = \frac{D_i}{L_L}(C_{LF,i} - C_{B,i}) \qquad (3.96)$$

where L_L is the thickness of the liquid boundary layer (L), D_i is the molecular diffusivity in water ($L^2 T^{-1}$), and $C_{B,i}$ and $C_{LF,i}$ are the concentrations of the component in the bulk liquid and at the biofilm surface (ML^{-3}), respectively.

3.4.3 Mathematical treatment with AQUASIM

The equations of the N1 model always have to be solved numerically, but various methods can be applied. For example, the partial differential equations can be converted by a finite-difference spatial discretization scheme (the method of lines), creating a system of algebraic and ordinary differential equations. For the time integration of this equation system, the fully implicit algorithm of Gear (1971) can be used, which was extended to differential-algebraic systems and implemented by Petzold (1983). The ability of the Gear-algorithm to handle a stiff set of equations is very important for the solution.

A convenient alternative is AQUASIM, a computer program designed for the identification and simulation of aquatic systems (Reichert 1998a, 1998b). The use of AQUASIM offers a number of features that are advantageous for simulations with the N1 model:

- Variables and processes readily can be activated or inactivated, making it simple to evaluate different model formulations.
- AQUASIM requires only input data on active processes (Section 3.4.1); all processes for which no data are provided automatically are inactive.
- If not stated otherwise, the biochemical and abiotic transformation reactions are automatically calculated for all compartments and phases of the system.
- The substratum can be selected to be flat, spherical, or cylindrical, and AQUASIM automatically adapts the mass balance equations accordingly.

- To model molecular diffusion in the biofilm, the values of the diffusivities in pure water can be used, since in AQUASIM diffusion in the biofilm is reduced by the assumption that it only takes place in the biofilm liquid phase.
- AQUASIM includes tools for data analysis by which experimental and calculated data can be compared, and sensitivity analyses and fits of the values of model parameters are performed.
- AQUASIM is based on robust numerical algorithms that, for most situations, calculate steady state and dynamic solutions without the need to adjust the numerical parameters.

More details on the numerical techniques used in AQUASIM and on the implementation concepts can be found in Reichert (1994). An example for the implementation of a heterotrophic-autotrophic biofilm is given in a tutorial (Reichert 1998b).

3.4.4 Applications

3.4.4.1 Substrate removal

AQUASIM can be used to model substrate removal in a biofilm reactor. Based on the kinetics of Benchmark 3 (BM3) described in Section 4.4, the reactor substrate outflow can be calculated as a function of the substrate inflow and the development of the biofilm in the reactor. The example in Figure 3.6 shows the substrate outflow decreasing during the first days because of biofilm growth. Then, after about three days, biofilm growth and biomass detachment reach a steady condition, and the substrate outflow remains constant. Figure 3.6 is an original plot as it is produced by AQUASIM.

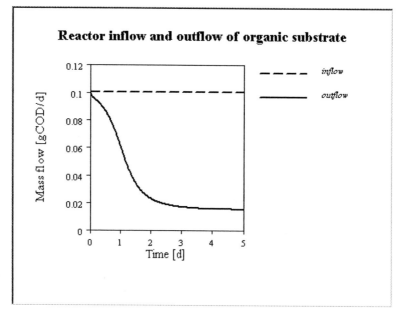

Figure 3.6. Typical AQUASIM plot showing the development in time of the reactor inflow and outflow of organic substrate.

3.4.4.2 Biofilm growth, microbial composition and detachment

AQUASIM can model biofilm growth as result of the production of microbial mass in the biofilm. Figure 3.7 shows an example of the development of the thickness and microbial composition of the biofilm. Again, this example is based on the kinetics of BM3 (Section 4.4). The inserts in Figure 3.7 display the relative abundance of autotrophic and heterotrophic microbial species and inert mass in the biofilm. In the beginning, the fast growing heterotrophic organisms dominate throughout the biofilm. After fifteen days, as the biofilm gets thicker, the slow-growing autotrophic microorganisms are more abundant in the biofilm depth, while the heterotrophic organisms still dominate near the biofilm surface. At 19.5 days, a sloughing event detaches most of the biomass. Autotrophic organisms dominate the remaining biomass. In AQUASIM, a sloughing event can be modeled as

$$u_{de} = 0.5\, u_F \qquad \text{for } t \le 19.5 \text{ days}$$
$$u_{de} = 500\, u_F \qquad \text{for } 19.51 \le t \le 19.52 \text{ days} \qquad (3.97)$$
$$u_{de} = 0.5\, u_F \qquad \text{for } t \ge 19.53 \text{ days}$$

where u_{de} is the global velocity of surface detachment (equation (3.93)) and u_F is the velocity by which the biofilm surface is displaced as a result of the production and decay of microbial mass in the biofilm (equation (3.89)). For most of the time, the detachment velocity u_{de} is smaller than the production velocity u_F, and the biofilm is growing. However, between 19.51 and 19.52 days, u_{de} has a value much larger than u_F, leading to an increased detachment of biomass to the bulk fluid and a rapid decrease of the biofilm thickness. Between 19.5 and 19.51 days and again between 19.52 and 19.53 days, the value of u_{de} is linearly interpolated. Many other possibilities to model sloughing are available in AQUASIM. One of them is to define a base thickness above which all biofilm is removed during a sloughing event (Morgenroth and Wilderer 2000; Horn *et al.* 2003).

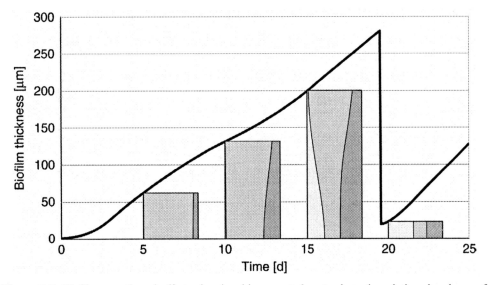

Figure 3.7. Biofilm growth and effect of a sloughing event. Inserts show the relative abundance of autotrophic (light gray) and heterotrophic (medium gray) microbial species and of inerts (dark gray) in the biofilm between the substratum (bottom) and the biofilm surface (top).

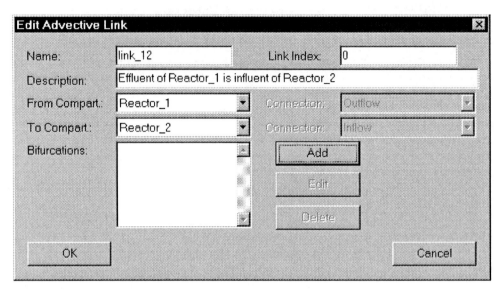

Figure 3.8. The AQUASIM dialog box "Edit Advective Link" is used to model water flow and advective substance transport from one compartment to another.

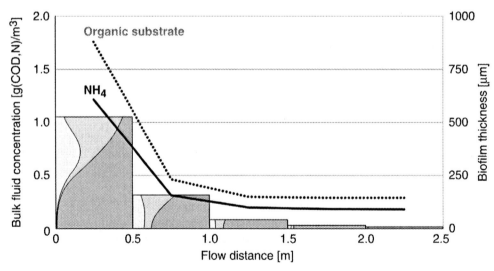

Figure 3.9. Pseudo 2d modeling by a series of five biofilm reactor compartments. In the reactor inflow, the concentration of organic substrate is 10 g_{COD}/m^3 and of ammonia 6 g_N/m^3. Inserts show the relative abundance of autotrophic (light gray) and heterotrophic (medium gray) microbial species and of inerts (dark gray) in the biofilm between the substratum (bottom) and the biofilm surface (top).

3.4.4.3 Pseudo 2d modeling of plug flow

A plug-flow biofilm reactor can be modeled in AQUASIM by a series of connected biofilm reactor compartments. The dialog box in Figure 3.8 shows an advective link, which is used to model advective mass fluxes between two reactor compartments. The effect of this link is that the effluent from Reactor 1 automatically becomes the influent to Reactor 2.

Figure 3.9 shows the biofilm in a plug-flow reactor modeled by 5 AQUASIM biofilm reactor compartments in series. In each compartment, the biofilm develops independently, and gradients of substrate concentrations and biofilm composition occur in the direction of flow. The biofilm that develops in the first compartment consists of a layer of inerts near the substratum, a layer of heterotrophic organisms near the biofilm surface, and a layer of autotrophic organisms between them. In the other compartments, the distribution of the microbial species is more or less uniform. In this biofilm reactor model, mass transfer in the flow direction occurs only between the bulk fluid zones of the compartments, while transport in the biofilm occurs only in the direction perpendicular to the substratum. Thus, the model is pseudo 2d. The kinetics for this example was also taken from BM3 (Section 4.4).

3.4.4.4 Pseudo 3d modeling

AQUASIM allows heterogeneous biofilm morphology to be taken into account by a simplified approach, based on the combination of multiple 1d simulations as implemented in the models with the codes NP3a, NP3b and NP3c. The basic idea for these pseudo-multidimensional simulations is the assumption that, even in heterogeneous biofilm morphologies, local mass transport is still mainly perpendicular to the substratum. Thus, the substratum area is divided into a number of sections, for which one-dimensional mass transport in the biofilm is calculated individually, but with the same bulk liquid substrate concentration. The overall performance of the biofilm is then evaluated based on a linear combination of the individual simulations.

The pseudo-3d models used in Chapter 4 for the benchmark calculations differ by the following characteristics: NP3a and NP3b are based on a steady state biofilm thickness distribution (Morgenroth *et al.* 2000), but in NP3a the bulk phase below the maximum biofilm thickness is assumed to be stagnant (Figure 3.10A), whereas in NP3b the boundary layer at the biofilm surface layer is neglected (Figure 3.10B). The latter assumption also holds for model NP3c, in which the detachment rate is not constant, but varies with time (Morgenroth and Wilderer 2000).

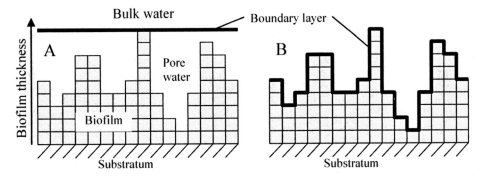

Figure 3.10. Modeling mass transfer across the surface of biofilms with heterogeneous morphology. The extreme assumptions are that mass transport below the maximum biofilm thickness is by diffusion only (A) and that the bulk liquid plus the biofilm pore water is completely mixed (B).

To calculate the overall flux for a steady state biofilm thickness distribution, simulations are consecutively performed for the individual biofilms on the various sections of the substratum area. Subsequently, the results from the individual simulations are combined using weights for each section. Thus, the overall specific flux $j_{F,S,tot}$ [$M_SL^{-2}T^{-1}$] of a component S can be calculated as

$$j_{F,S,tot} = \sum (j_{F,S,i} w_i)$$ (3.98)

where $j_{F,S,i}$ and w_i are the flux [$M_SL^{-2}T^{-1}$] and the weight [-], respectively, associated with a specific biofilm thickness $L_{F,i}$ [L]. If the sum of all w_i is one, the average biofilm thickness $L_{F,avg}$ [L] equals

$$L_{F,avg} = \sum (L_{F,i} w_i)$$ (3.99)

where $L_{F,i}$ is the biofilm thickness obtained by simulation i. All simulations can be performed independently. However, they are related to each other by the common bulk liquid concentration $S_{B,S}$ [M_SL^{-3}], which is determined as

$$S_{B,S} = S_{in,S} - \frac{A_F j_{F,S,tot}}{Q}$$ (3.100)

where $S_{in,S}$ is the influent concentration of S [M_SL^{-3}], A_F is the total biofilm area [L^2], and Q is the influent flow rate [L^3T^{-1}]. The calculation of the overall flux has to be done iteratively: First, a value of the bulk liquid concentration is assumed. With this value, the simulations are performed for all sections, and $j_{F,S,tot}$ and $S_{B,S}$ are calculated by equations (3.98) and (3.100), respectively. Then, the calculated and the assumed $S_{B,S}$ are compared, and the simulations are performed again with a corrected bulk liquid concentration. This procedure is repeated until the assumed and the calculated $S_{B,S}$ are equal.

The overall flux for a biofilm that is subject to local dynamic detachment also can be approximated by 1d simulation. Detachment events in a biofilm can be expected to occur randomly distributed over the entire biofilm surface and, in a steady state situation, to result in a constant average biofilm thickness $L_{F,avg}$ (Figure 3.11). In model NP3c, the underlying assumption is that local detachment events occur when the biofilm at this location reaches a certain thickness (Morgenroth and Wilderer 2000). Reduction of the biofilm thickness down to a predefined base value occurs as sudden sloughing in regular intervals, where the durations of the intervals are selected such that a certain time average of the biofilm thickness can be maintained. In NP3c, the overall specific flux $j_{F,S,tot}$ [$ML^{-2}T^{-1}$] of the component S across the biofilm surface is calculated by averaging fluxes over time as

$$j_{F,S,tot} = \frac{\int_{t_1}^{t_2} j_{F,S}(t)dt}{t_2 - t_1}$$ (3.101)

where $j_{F,S}(t)$ is the flux at time t [$ML^{-2}T^{-1}$], and t_1 and t_2 refer to the occurrence of two subsequent detachment events [T]. Again, the problem has to be solved iteratively, by the same procedure as described in the last paragraph.

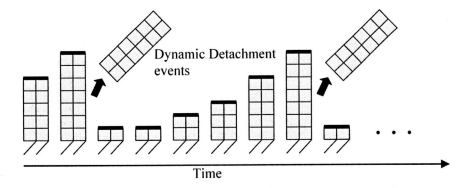

Figure 3.11. Conceptual model of local dynamic detachment events for biofilms with heterogeneous morphology.

3.5 NUMERICAL ONE-DIMENSIONAL STEADY STATE MODEL (N1s)

3.5.1 Features

Analytical and pseudo-analytical models like the ones discussed above approximate the steady state variant of the nonlinear diffusion-reaction equation (3.20) by linear problems that are easier to solve. The reason for this is that no analytical solution to the actual nonlinear model can be found. The idea behind the analytical and pseudo-analytical methods is to approximate the more complete, but not explicitly solvable nonlinear model by a simplified and explicitly solvable model. On the other hand, accurate numerical methods are available to solve the full nonlinear equation (3.20) approximately. Pursuing this avenue, the flux (3.53) with (3.51), (3.52) is replaced by a numerical approximation. Although it is possible in theory to implement such a model in a spreadsheet, it is not advisable due to arithmetic inaccuracies and memory inefficiencies of those software products. Instead, special scientific-technical software (see section 3.5.3) should be used. The nonlinear equation (3.20) is embedded in the global mass balance (3.43) of the reactor.

The convenience of not having to track the dynamic evolution of the biofilm is in some cases at the expense of additional model assumptions that must be introduced, based on the intuition and experience of the modeler. The same situation was already encountered in the context of the analytical and pseudo-analytical model and it can be dealt with in the same manner. For example, in multi-species problems *a priori* information about the relative spatial distribution of the microorganisms must be provided. In the application of the model to BM3 below the assumption is made that the three particulate fractions heterotrophs, autotrophs and inerts are all homogeneously distributed in the biofilm, i.e. at every point in the biofilm all three particulate substances are present and their concentrations are constant over the entire biofilm. An alternative would be to assume the layer of inerts by the substratum and the fastest growing group of bacteria at the biofilm/liquid interface. In the first case the constant biomass concentrations for all particulate fractions are computed as an output variable. In the second case it is assumed that in the individual layer the maximum biomass density is attained throughout and the thickness of the layers must be computed. The shortcoming that additional model assumptions must be made, of course, has a positive side to it as well, because the user is required to carefully analyze the system to be modeled and to review each model assumption individually. The big advantage of the modeling concept is its flexibility. Among the 1d models covered in this survey, model N1s is the only one that could be used for a predictive solution for all three benchmark problems in Chapter 4.

In all cases the model reduces to a nonlinear system of equations for only few variables. In order to evaluate these functions, however, the nonlinear system of ordinary differential equations describing diffusion and reaction processes inside the biofilm must be solved. Since in general no analytical solutions for these differential equations can be found, they are treated numerically. A variety of numerical methods can be used. These problems are not critical with respect to memory or computing time for regular personal computers.

3.5.2 Definitions and equations

The basic equations are the reactor mass balance (3.44), the transport-reaction equation in the biofilm (3.20) and the flux condition (3.25) that connects both compartments. In order to be able to solve (3.20), also the distribution of biomass in the biofilm must be known. In some cases this will be given as input variable (cf. Benchmark Problems 1 and 2) in other cases it has to be computed, sometimes with the help of additional equations and/or simplifying model assumptions (cf. Benchmark Problem 3). A similar statement can be made for the operating conditions of the reactor, in particular with the respect to hydrodynamic parameters as in Benchmark Problem 2. An example will be given below.

It should be noted that the global mass balances can also be used in a straightforward manner to determine model parameters, such as the biofilm thickness or the diffusion coefficients from concentration measurements in the reactor by solving an inverse problem.

3.5.3 Software Implementation

Steady-state mass balances of this type reduce to small systems of nonlinear equations (with only few components) at the core of which ordinary 2nd order two-point boundary value problems must be solved. This type of numerical problem can be easily implemented using numerical/mathematical software packages (such as Octave *(free)*, SciLab *(free)*, Matlab *(commercial)*, Mathematica *(commercial)*, MAPLE *(commercial)* to name but a few) or programming languages such as Fortran, C/C++ and Java. Many of these software environments offer efficient and extremely accurate ready to use routines for the individual modules, which greatly simplifies the implementation. While the numerical/mathematical software packages typically have a built-in graphical output capability, special visualization software such as Gnuplot *(free)* or AVS/Express *(commercial)* can be used to visualize results computed with self-written programs.

A software implementation that is ready to use is currently not available. This is in part due to the observation that every biofilm system has its own peculiarity and it might be necessary to introduce system specific additional sensible simplifications or assumptions of the quantities involved (see the applications below). In order to obtain a closed and well-posed model a sound understanding of the biofilm processes and in particular of their inter-dependencies is required. Of course, this has its advantage as well, because the modeler is forced to thoroughly think through her/his problem. Hence, some familiarity with mathematical software packages or programming experience is required.

3.6 MULTI-DIMENSIONAL NUMERICAL MODELS (N2 and N3)

The increasing amount of experimental evidence regarding biofilm heterogeneity has driven the development of more complex biofilm models intended to describe some of the heterogeneous characteristics of biofilms. The premise is that, by capturing the spatial and temporal heterogeneity of the physical, chemical, and biological environment, the model makes it possible to obtain an assessment of biofilm activity and interactions at the microscale. New problems to be addressed by multi-dimensional (multi-d) biofilm models include, for example:

- *Geometrical structure of biofilms*: How does the spatial biofilm structure form? What is the influence of environmental conditions on the biofilm structure? How does quorum sensing operate? What causes biomass detachment? How does microbial motility influence biofilm formation?

- *Mass transfer and hydrodynamics in biofilms*: What is the importance of advective mass transport relative to diffusion in the biofilm? How does the biofilm's spatial structure affect the overall solute transport rates to/from the biofilm?

- *Microbial distribution in biofilms*: What is the importance of inter-species substrate transfer? What is the influence of substrate gradients on microbial competition and selection processes?

The objective of this chapter is to provide a general overview and comparison of several multidimensional (2d and 3d) biofilm models used in the benchmark comparisons from Chapter 4. Several other approaches are described in the literature, but their extensive presentation is beyond the scope of this report (see reviews in Picioreanu *et al.* 2000c, 2003, 2004).

3.6.1 General features

In 2d and 3d models, some of the assumptions used by 1d models are relaxed to allow additional processes to be considered. For instance, the assumptions of steady-state used in the development of analytical (the A model, Section 3.2) and pseudo-analytical (PA models, Section 3.3) solutions to the biofilm problem or the concepts of uniform thickness, layering of biomass, or completely mixed bulk liquid commonly used in 1d analyses (N1s) are not necessarily maintained. All the multi-dimensional models are based on the same physical and biological principles of mass transport and substrate utilization generally used in the dynamic 1d models (such as N1), but extended to 2 or 3 spatial dimensions.

Similar to 1d models, the domain in which a biofilm develops is composed of two regions: the *biofilm* (with biomass comprising EPS and cells) and the *liquid*. A 1d situation has a clear separation between the biomass and liquid regions. When the problem is expanded to a 2d or 3d domain, the definition of the two regions may become less clear (see Figure 2.2). For instance, channels and pores within the biofilm might develop, and the interface between biomass and liquid regions might be highly irregular (Figures 2.4 and 2.5).

The main difference between 1d and multi-d models is in the way processes affecting the development of the solid biofilm matrix and the dynamics of its composition (i.e., biomass growth, decay, detachment and attachment) are simulated. For example, when second or third dimensions are part of the physical domain being modeled, the biofilm matrix has more than one direction in which to grow, allowing the simulation of spatially heterogeneous biofilms (Picioreanu *et al.* 1998a; Noguera *et al.* 1999b; Eberl *et al.* 2001; Laspidou and Rittmann 2004; Alpkvist 2005). While some models assume that the biofilm density is constant in the entire biofilm structure (Dockery and Klapper 2001; Hermanowicz 2001;

Mehl 2001), other models treat it as an *a priori* unknown variable, governed by evolution equations (Picioreanu *et al.* 1998a; Eberl *et al.* 2001; Kreft *et al.* 2001; Pizarro *et al.* 2001; Eberl and Efendiev 2003; Noguera *et al.* 2004; Laspidou and Rittmann 2004). The latter approach has been applied to describe also the interaction of different biomass fractions in a multispecies biofilm, while up to now, biofilm models using an *a priori* defined constant biomass density seem to be restricted to situations with only one biomass fraction.

Other potentially important phenomena in the analysis of the overall biofilm system include the effect of fluid motion and advective transport of substrate in and out of the biofilm. This has been studied by coupling multidimensional biofilm models with fluid dynamics modeling of the liquid phase environment (Picioreanu *et al.* 2000a; Eberl *et al.* 2000a; Mehl 2001).

Another class of addressed problems concerns the interaction among biofilm shape, fluid flow, biomass decay, and biofilm detachment (Picioreanu *et al.* 2001; Pizarro *et al.* 2001; Noguera *et al.* 2004). Different approaches to biofilm detachment have been proposed, the most important being: breaking caused by biomass decay (Pizarro *et al.* 2001; Noguera *et al.* 2004), erosion and sloughing due to mechanical stress induced by fluid flow (Picioreanu *et al.* 2001), and using general detachment rates (Xavier et al. 2005a, 2005b, 2005c).

Of course, allowing more degrees of freedom in the models causes the complexity of the model to increase dramatically, and the computing requirements may easily become limiting. However, although initially some of the 2d and 3d models were coded and run using high performance supercomputers (Picioreanu 1999; Eberl *et al.* 2000b), nowadays, most multidimensional biofilm models can be executed on single-processor machines (e.g., PCs) (Xavier *et al.* 2005a), with the exceptions of those 3d models that explicitly incorporate hydrodynamics.

3.6.2 Model classifications

3.6.2.1 Definitions

Multi-dimensional biofilm models generally treat two distinct regions: the *biofilm* phase (with biomass comprising EPS and cells), and the *liquid* phase in which the biofilm develops. As for the N1 model, two fundamental types of variables are distinguished, which are treated differently. The *particulate* components form the biofilm solid matrix and the *dissolved* components exist in the biofilm phase and in the bulk liquid. The N1 type of model assumes the existence of a representative element of volume (REV) over which an average of any relevant propriety (e.g., biofilm density, porosity, or every dissolved and particulate component concentrations) is made. Consequently, all points within any REV the state variables have the same value (i.e., the REV is fully homogeneous). [A comprehensive analysis of conditions under which biomass averaging is a valid computational tool was made by Wood and Whitaker (1998, 1999)]. The REV should be much larger than the size of a bacterium, but also larger than the typical distance between bacteria. However, REV size must be small compared to the characteristic length scale over which biomass significantly changes in order to have a good spatial resolution of the biomass distribution in the biofilm. Typical REV sizes are in the order of tens of microns.

3.6.2.2 Representation of dissolved components

The assumption that sections of the biofilm can have a homogeneous composition leads to the definition of a *continuum* space for some model variables. This is ordinarily the case with the dissolved components when we are interested only in the concentration of a certain component (e.g., mg/L dissolved oxygen) and not the dynamics of each molecule. Due to the continuous space, mass balances can be written as differential equations and, thus, the powerful apparatus of differential calculus and numerical analysis can be applied for their solution. 2d and 3d transport processes such as diffusion and advection can be easily integrated with conversion processes in mass balances of type given by the general equation (3.3) or the subsequent particular forms (3.9) to (3.20). Models N2a, N2b, N3a, N3b and N3c used for the benchmark problems fall into this category.

An alternative to solving the mass balances of dissolved components presenting the advantage of simplicity is using *discrete-stochastic* methods such as cellular automata (CA). In this case, solutes are modeled as discrete particles randomly moving in the 2d or 3d domain, and substrate gradients are the result of localized differences in the number of substrate particles within the computational domain. Nevertheless, the algorithms used for the random walks of particles are such that on average one obtains the same concentrations as those resulted from continuum based physical representations. Models N2c, N2d, N2e and N2f use discrete CA methods for the representation of dissolved components.

3.6.2.3 Representation of particulate components

An essential aspect that must be considered when building any biofilm model is the foreseen spatial scale (spatial resolution). No consensus exists on which approach is the best, and the "bestness" depends on particular conditions. Current models for biofilm structure deal in two different ways with bacteria, depending mostly on the biofilm scale targeted. The first class of models treats biomass as a *continuum* (Figure 3.12A), and biomass transport (spreading) is generally achieved by applying differential equations widely used in mechanics and transport phenomena. To obtain the dynamics of spatial distribution of biomass, one solves partial differential equations of form (3.3). Although currently in development (Eberl *et al.* 2001; Dockery and Klapper 2001; Alpkvist 2005), this class of biofilm models was not used for the multi-d approach to benchmark problems in this report.

In the second approach, instead of solving equation (3.3) in a continuum medium, the biomass balance is split into two separate processes, treating transport and reaction differently. While the dynamics of biomass accumulation is still governed by an ordinary differential equation (e.g., biomass growth kinetics), the transport part is realized with *discrete* biomass units. Furthermore, the discrete biomass spreading methods can be divided in two classes. In grid-based methods (including cellular automata algorithms, CA), transport of biomass is only stepwise along a finite number of directions according to a set of discrete rules (Figure 3.12B). Such models can use volume averaging to develop macroscopic equations for biomass evolution and the mass of cells per unit volume (density or concentration, X) is the biomass-related state variable, or stochastic rules to estimate when a biomass particle with a fixed biomass density will multiply producing an identical daughter particle. Models N3a, N3b, N2c, N2d, N2e and N2f use a grid-based discrete biomass representation with CA implementations. The second line of discrete models is based on biomass particles (e.g., individual based-modeling, IbM). Particle-based models allow biomass movement on a continuous set of directions and distances (Figure 3.12C). IbM allows individual variability and may treat bacterial cells as the fundamental entities. Essential state variables are, for example, the cell mass m, cell volume V, etc. In this category is the N2a model.

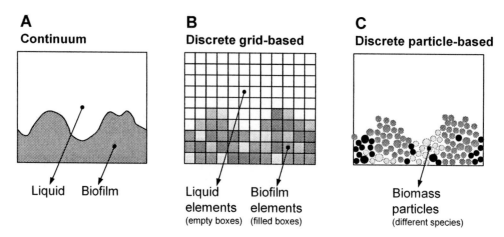

Figure 3.12. Multi-d biofilm models can treat biomass (A) as a continuum, (B) as discrete elements on a grid (e.g., cellular automata - CA), or (C) as particles (e.g., individual-based models - IbM).

A large number of CA-based biofilm models has been published in recent years (Wimpenny and Colasanti 1997; Hermanowicz 1998; Picioreanu *et al.* 1998a, 1998b; Noguera *et al.* 1999b; Gonpot *et al.* 2000; Hermanowicz 2001; Picioreanu *et al.* 2001; Pizarro *et al.* 2001; Chang *et al.* 2003; Hunt *et al.* 2003; Laspidou and Rittmann 2004).

A variety of discrete rules, although easy to implement, may produce qualitatively different and rather arbitrary structures. They might easily mislead the researcher to aesthetically driven rather than physically motivated model formulation. Eberl *et al.* (2001) pointed out a number of inherent drawbacks that discrete/stochastic mathematical models for biofilm spreading possess: (1) Due to the small set of possible directions for movement, these models are not invariant to changes of the coordinate system; (2) *Ad hoc* rules for priority must be specified for simultaneous shifts into the same grid cell; (3) Many different and somewhat arbitrary rules can be formulated for spreading; (4) Complicated stochastic (branching) processes and one simulation is only one realisation of this stochastic process. In addition, we have seen from numerous computational experiments (e.g., Kreft *et al.* 2001) that, when applied to multispecies biofilms, the CA-based model presented in (Picioreanu et al. 1998a, 1998b) seem to produce too much internal mixing (dispersion) of species within a colony. This is a direct consequence of the stochastic pushing rules used for biomass redistribution. Multi-species biofilm simulations with a variant of the discrete algorithm that eliminates this problem tend to generate anisotropic colonies instead (Picioreanu *et al.* 2000c). Noguera *et al.* (1999b) used a recursive algorithm to find appropriate locations for the spreading of excess biomass formed during a time step. If all the elements adjacent to the element where biomass growth took place were fully occupied with biofilm, the model searched for the closest non-adjacent available element and then, the biomass was pushed from the source element towards the available element. This biomass-spreading method was successfully used in the simulation of multispecies biofilms. A newer version of cellular automaton model (Pizarro *et al.* 2001) uses "microbial particles" of certain mass. The model can be used to simulate biofilms of uniform biomass density simply by redistributing the microbial particles at the end of each iteration (Pizarro *et al.* 2001), or it can be used to simulate the formation of heterogeneous biofilms by excluding such redistribution rule (Noguera *et al.* 2004).

3.6.2.4 Summary of multidimensional models used

Table 3.10 summarizes the multi-d models used in this report, together with the main processes included and with further literature references. The next sections describe each model in more detail. All multidimensional models are using numerical solutions of soluble and particulate concentration fields.

Table 3.10. Biofilm models applied in the benchmarks

Code	Specification	Benchmark	Reference
N3a	3d model with differential single solute diffusion and reaction, differential biomass growth and decay, and cellular automata biomass spreading. Mass transfer boundary layer is parallel with substratum (see Figure 2.2B).	BM1	Section 0 Picioreanu et al. (1998a, 1998b, 1999)
N3b	As N3a, but mass transfer boundary layer follows the biofilm surface (liquid pores completely filled, see Figure 2.2C).	BM1	Section 0 Picioreanu et al. (1998a, 1998b, 1999)
N3c	A fully 3d numerical simulation of hydrodynamics and mass transfer in the specified biofilm architectures (the original problem; serves as reference for comparison)	BM2	Section 0 Eberl et al (2000a)
N2a	2d reduction of a 3d model including differential diffusion and reaction of multiple solutes, differential growth and decay, individual-based transport and detachment of multiple biomasses.	BM3	Section 0 Picioreanu et al. (2001, 2004); Xavier et al. (2005a, 2005b, 2005c)
N2b	2d model with full differential hydrodynamics, differential single solute diffusion, advection and reaction, differential biomass growth and decay, and cellular automata biomass spreading.	BM2	Sections 0 and section 4.3.3.2.1 Picioreanu et al. (2000a, 2000b, 2001)
N2c	2d model with cellular automata solute diffusion and reaction, and cellular automata biomass growth, decay and spreading.	BM1	Section 3.6.4 Pizarro et al. (2001); Noguera et al. (2004)
N2d N2e	As in N2c, plus further simplified cellular automata 1d hydrodynamics.	BM2	Section 3.6.4.6 Noguera et al. (2004)
N2f	As in N2c with multiple biomass components	BM3	Section 3.6.4.5 Pizarro et al. (2001); Noguera et al. (2004)

3.6.3 2d and 3d models with discrete biomass and solutes in continuum space (N2a, N2b, N3a, N3b, N3c)

3.6.3.1 Features

This section describes the general features of the series of multi-d biofilm models N2a, N2b, N3a, N3b and N3c used in the benchmarks performed for this report. The models N2a, N3a, N3b consider: (1) any number of dissolved components distributed in a continuous medium and (2) any number of particulate components represented by discrete biomass entities.

The outputs produced by these models include the time-dependent
- 3d spatial distribution of any number of particulate components (e.g., biomass, EPS, inert material) in the biofilm
- architectural properties of biofilms, such as the surface shape and porosity
- accumulation and loss from the system of the mass of particulate components, including detachment rates
- 3d spatial distribution (fields) of concentrations of any number of dissolved components (e.g., substrates, products)
- local and overall removal rates of dissolved components
- fluid flow patterns past the biofilm (e.g., fluid velocity, shear stress on the biofilm surface) - models N2b and N3c only
- local and overall parameters of the external mass transfer of dissolved components

The processes considered include
- any transformations of dissolved and particulate components, such as substrate uptake, biomass growth, decay, and EPS production
- spreading of particulate components in the biofilm due to volume-generating biomass transformations, realized by particle divisions and particle pushing (i.e., empirical advection and diffusion rules)
- detachment and attachment of particulate components at the biofilm surface
- diffusion and advection of dissolved components in the mass transfer boundary layer and diffusion only in the biofilm phase
- complete mixing of dissolved components in the bulk liquid

The data required by these models consist of
- diffusion coefficients in pure water and in the biofilm for all dissolved components (can be variable in space)
- inflow concentrations of dissolved components, inflow volumetric flowrate of liquid to the biofilm system, biofilm substratum area, bulk liquid volume and boundary layer thickness – when bulk concentrations are calculated from mass balances (N2a, N3a, N3b), or
  inflow concentrations of dissolved components and liquid velocity profiles – when concentrations in the boundary layer are calculated from mass and momentum balances (hydrodynamic models, N2b, N3c)
- the stoichiometry (coefficients) and kinetics (rate equations and parameters) of all transformation processes
- density of all particulate components and maximum mass (or size) of biomass particles at division (for IbM)
- detachment functions of, e.g., biofilm depth, biofilm composition, biofilm shape – for generic detachment models, or

mechanical biofilm properties, e.g., elasticity modulus, maximum stress – for detachment models based on structural mechanics
- attachment rates, if this is considered
- the geometry of the biofilm substratum (planar, spherical, or any irregular surface)
- initial distribution of particulate components on the substratum, i.e., number, size, type and composition of the biomass particles attached to the substratum at the beginning of the simulation

The current applications of the multi-d models target primarily diverse areas of research and education. Like N1 models, multi-d models can be used to simulate behavior of time-dependent and steady-state variables in multi-species and multi-substrate biofilms. In microbiology and microbial ecology research, these models can be used as tools to express hypotheses about formation of biofilm structure and about microbial interactions in complex communities. They can be used to describe and to evaluate experimental data. In principle, multi-d models are applicable also in engineering applications, but (at least for now) the high spatial resolution and level of detail do not match well with engineering goals, which are more macroscale.

Limitations of multidimensional models include:
- inherent higher complexity than 1d models (e.g., N1) at computational and understanding levels
- no standardized way to describe the biofilm system; instead many approaches are being taken
- the many different computer software implementations are not freely available (with a few exceptions, see Section 3.6.3.4)
- still in the phase of development in universities, but not in commercial applications

The time evolution of concentrations of dissolved components is computed numerically from non-linear second-order partial differential mass balance equations. The evolution of particulate components and the biofilm geometrical structure result from ordinary differential equations for mass change and from empirical (discrete-stochastic) models for biomass spreading.

3.6.3.2 Definitions and equations

The series of discrete-continuous multi-d biofilm models used in this report were described in detail in several publications in recent years (Picioreanu *et al.* 1998a, 1998b, 2000a, 2000b, 2001, 2004; Eberl *et al.* 2000a; Xavier *et al.* 2005a, 2005b, 2005c). This section contains a brief description of the main definitions, equations, and algorithms used.

3.6.3.2.1 Biofilm system definition

The biofilm system is defined in different ways in the models presented here. The biofilm grows in a rectangular domain (box) with dimensions $L_X \times L_Y \times L_Z$, on a planar, inert substratum with the surface at $z=0$. In 2d, the domain has size $L_X \times L_Z$.

(a) In models N2a, N3a, and N3b used for BM1 and BM3, the biofilm system contains two domains: (1) the completely mixed domain, and (2) the diffusion domain (Figure 3.13a). The bulk liquid volume V_B (m³) is very large compared with the biofilm volume V_F (m³), and, thus, V_B can be considered constant.
- The *completely mixed domain* B is equivalent to the bulk liquid compartment in the N1 models. The bulk liquid is continuously fed at a flow rate Q (L³T⁻¹) (which can be variable in time) with a solution containing a number of n_S soluble substrates i in concentrations $S_{in,i}$

($M_S L^{-3}$) (which can also be variable in time). Liquid is also continuously withdrawn from the reactor at the same flow rate Q. The substrate concentration in the bulk liquid is $S_{B,i}$.

For simplicity, biological activity and other reactions are neglected in the bulk liquid. Biomass is not contained in the inflow.

• The *diffusion domain* contains in turn two sub-domains: a biofilm region F and a mass-transfer boundary layer L (MTBL). The biofilm develops on a planar support with area A_S. Because of growth, decay, and detachment, the volume V_F of the biofilm region changes in time. Dissolved components are transported by diffusion in the boundary layer and biofilm. For simplicity, transformation processes (reactions) are considered only in the biofilm region. The shape of the biofilm/BL interface (i.e., the biofilm surface) is irregular and continuously changes in time (i.e., moving boundary problem). Consequently, the position and surface area A_F of the biofilm surface changes. The shape of the BL/bulk interface can be set either as a plane parallel to the substratum (case model N3a) or following the biofilm surface at a certain distance (case model N3b, see also Figure 3.14). The computational domain is only a small part of the whole system, including a small fraction of bulk liquid and biofilm.

(b) In the models N2b and N3c, used for BM2, the biofilm system contains two different domains: (1) a liquid region L and (2) a biofilm region F separated by an interface Γ (Figure 3.13b).

• In the *liquid domain*, liquid flow is driven by moving the top of the domain with a constant velocity, $u_{X,max}$. Values can be prescribed for the liquid velocity $\mathbf{u}$ and for the n_S concentrations of dissolved components $S_{in,i}$ in the liquid inflow. Dissolved components are transported by diffusion and advection. By applying consistent boundary conditions, the 2d or 3d spatial distributions of liquid velocity $\mathbf{u}(x, y, z)$ and concentrations $S_{L,i}(x, y, z)$ in the whole liquid domain can be calculated by computational fluid dynamics methods. Again, for simplicity only, transformations are neglected in the liquid domain, and biomass is not contained in the inflow.

• The *biofilm domain* is treated as a solid body through which dissolved components can only diffuse. Transformation processes are due to the particulate components present (i.e., biomass), leading to formation of new biomass and changes in the biofilm volume.

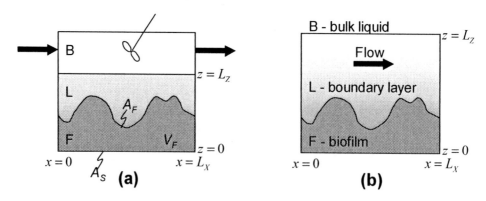

Figure 3.13. Schematic representation of model biofilm system and computational domains in (a) models N2a, N3a, and N3b; (b) models N2b and N3c.

3.6.3.2.2 Mass balances for dissolved components

The spatial distribution ("the field") of the concentrations of a given set of dissolved components (substrates, products, etc.) influences the growth rate of a particular biomass type. Conversely, the spatial distribution of bacterial activity affects the substrate concentration fields. Due to transformations (i.e., growth, division, decay), transport (i.e., spreading), and transfer processes (i.e., detachment, attachment), the spatial distribution of biomass varies in time. This causes a temporal variation of the concentration fields for soluble components as well. For this reason, mass-balance equations of time-dependent transport and transformation must be written for each dissolved component i ($i = 1,2,...,n_S$) in each domain of the computational biofilm system in order to find solute concentrations S_i.

1. Mass balances of dissolved components in the bulk liquid (B). The computational domain is only a small region in a larger environment. The processes taking place in the biofilm affect this environment, called the bulk liquid. In turn, solute concentrations S_B in the bulk liquid (presumably at $z>L_Z$) form boundary conditions for mass balances in the computational domain. Therefore, for each dissolved substrate, a mass balance equation in dynamic conditions is needed for each of the n_S solutes i (similar to equation (3.39)),

$$V_B \frac{dS_{B,i}}{dt} = Q\left(S_{in,i} - S_{B,i}\right) + \frac{A_F}{L_Y L_Z} \int_{V_F} r_{F,i} \cdot dv \qquad (3.102)$$

What equation (3.102) simply accounts for is that the accumulation of dissolved matter in the completely mixed bulk volume V_B (L^3) is the result of: (1) flow into and out of the bioreactor, with Q the volumetric flow rate of liquid through the bulk (L^{3}T^{-1}) and S_{in} the influent concentration, and (2) the global conversion rates in the biofilm volume V_F (L^3), with A_F the total biofilm surface area in the environment (L^2). The factor in (3.102) scales the substrate conversions obtained in the small "representative" computational volume (having a biofilm area $L_Y L_Z$) to the conversions achieved in the whole system (with biofilm area, A_F). In other words, $A_F/(L_Y L_Z)$ represents the number of times the biofilm area used in computations will fit within the total biofilm surface in the bioreactor. r_F is the net transformation rate of the component i in a very small biofilm volume element dv. Equation (3.102) needs initial conditions for the concentrations S_B at $t=0$.

Because mass balances in the bulk liquid depend on concentrations in the biofilm, new mass balances are needed for the biofilm domain. The solution of equations (3.102) is then used as boundary condition for the mass balances in the biofilm.

2. Mass balances of dissolved components in the biofilm (F) and boundary layer (L). Spatially multi-d models (i.e., 2d or 3d) relax the usual assumption made by 1d models that the direction of mass fluxes is only perpendicular to the substratum. This means that the general problem can now be described by mass balances including the mass flux **j** (ML^{-2}T^{-1}) change in two or three space directions, as given by the equation (3.3) applied to dissolved components:

$$\frac{\partial S}{\partial t} \quad = \quad -\frac{\partial j_x}{\partial x} - \frac{\partial j_y}{\partial y} - \frac{\partial j_z}{\partial z} \quad + \quad r \qquad (3.3)$$

As for 1d models, the net reaction term present in mass balance (3.3), r, is dependent on multiple dissolved substrate concentrations $\mathbf{S}=[S_1,...,S_{ns}]$ and biomass concentrations $\mathbf{X}=[X_1,...,X_{nx}]$:

$$r \quad = \quad r(\mathbf{S}, \mathbf{X}) \qquad (3.103)$$

For a chemical species i, the net transformation rate r_i ($ML^{-3}T^{-1}$) is obtained from the sum of the rates ρ_j of all the n_R reactions in which this species is involved, each multiplied by a stoichiometric coefficient v_{ij} specific to that reaction

$$r_i = \sum_{j=1}^{n_R} v_{ij}\rho_j \tag{3.104}$$

The preferred way to represent the stoichiometry is in a matrix form (see section 2.4.1), similar to the activated sludge models (ASM) description (Henze *et al.* 2000).

In the general case, transport of an electrically charged dissolved chemical species in a dilute solution is modeled by Nernst-Plank flux equations including molecular diffusion, migration in an electrical field and convection due to fluid flow (see equation (2.16)):

$$\mathbf{j} = -D\nabla S - \zeta DS\nabla\tilde{\Phi} + \mathbf{u}S \tag{3.105}$$

where S is the concentration of the dissolved substrate (ML^{-3}), $\tilde{\Phi}$ is a dimensionless potential function introduced by the necessity to maintain electroneutrality in each point in space, $\mathbf{u}$ is the vector of fluid velocity (LT^{-1}), D is the molecular diffusion coefficient (L^2T^{-1}) and ζ is the ion charge. Substituting equation (3.105) into (3.3) yields a general mass balance equation that describes the development over time and the spatial field of the concentration S of any dissolved component i as:

$$\frac{\partial S_i}{\partial t} + \mathbf{u}\nabla S_i = \nabla(D_i\nabla S_i) + \zeta_i D_i\nabla(S_i\nabla\tilde{\Phi}) + r_i(\mathbf{S},\mathbf{X}) \tag{3.106}$$

To be solvable for $[S_1,...,S_{ns}, \tilde{\Phi}]$, the system of nonlinear partial differential equations (3.106) has to be complemented with an electroneutrality condition in each point in space:

$$\sum_{i=1}^{n_S} \zeta_i S_i = 0 \tag{3.107}$$

This general situation appears when charged solutes (ions) are transported at different rates (e.g., different diffusion coefficients). Solving equations (3.106) and (3.107) is needed, for example, when pH gradients have to be calculated at low ionic strength of solution. Examples of how to deal with this situation can be found in a 1d carbon-limited algal biofilm model (Flora *et al.* 1993), the 3d bio-corrosion model (Picioreanu and Van Loosdrecht 2002), and in a model for formation of anaerobic digestion granules (Picioreanu *et al.* 2004a).

The usual simplification is to neglect ion migration, when either there are only neutral species or at high ionic strength, which transforms (3.106) into (3.108). Models N2b and N3c, used for BM2, apply mass balance (3.108) for dissolved components in the liquid domain L with the assumption that no reaction occurs in this domain:

$$\frac{\partial S_{L,i}}{\partial t} + \mathbf{u}\nabla S_{L,i} = \nabla(D_{L,i}\nabla S_{L,i}) \tag{3.108}$$

Solute advection in the boundary layer is not considered in models N3a, N3b and N2a, therefore equation (3.109) is used in these models for the boundary layer domain:

$$\frac{\partial S_{L,i}}{\partial t} = \nabla(D_{L,i}\nabla S_{L,i}) \tag{3.109}$$

Because advective transport of substrates inside the biofilm is typically negligible but various reactions can take place, one obtains in the solid biofilm region the diffusion-reaction equation:

$$\frac{\partial S_{F,i}}{\partial t} = \nabla(D_{F,i}\nabla S_{F,i}) + r_{F,i} \tag{3.110}$$

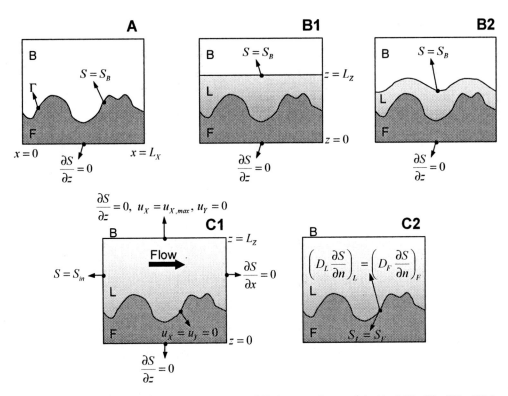

Figure 3.14. Boundary conditions used by: (A,B) diffusion-reaction models (A - N3b, B1 - N3a, N2a); (C) diffusion-convection-reaction models (C - N3c and N2b). B is the bulk liquid, L is the mass transfer boundary layer and F is the biofilm region.

Models N3a and N3b (for BM1), N2b and N3c (for BM2), and N2a (for BM3) use the mass balance (3.110) for dissolved components in the biofilm domain (see also Figure 3.3a). Because of time scale separation possibility (section 2.6.3), only the steady state ($\partial S/\partial t=0$) forms of equations (3.108)-(3.110) are solved (see solution methods in section 3.6.3.3.1).

The diffusion coefficients D_F inside the solid biofilm domain are space-dependent, are a function of the local biomass densities, and are normally smaller than the diffusion coefficients D_L in water (Beuling *et al.* 2000). For small molecules (like oxygen, ammonium or nitrate), $D_F=D_L$ often can be assumed a first approximation. For large molecules, $D_F=D_L f(X)$ can be assumed, with the effective D_F being as much as one order of magnitude lower than D_L (Bryers and Drummond 1998).

3. Boundary conditions. The mass-balance equations given here describe the processes inside the model domain and must be completed by appropriate boundary conditions describing the experimental setup or the modeled case. At the biofilm-support interface, $z = 0$, zero-flux conditions ($\partial S/\partial z = 0$) apply for all N2/N3 models. Other conditions, such as substrate permeability, are also applicable (e.g., in case of membrane biofilm reactors). Models using only diffusion-reaction substrate balances in form of (3.110) can assume, for simplicity, that the bulk liquid concentration $S = S_B$ (Figure 3.14A, model N3b) at the biofilm-liquid interface Γ. The other domain borders are sometimes (models N2a, N3a, N3b) wrapped-around in so-called periodic boundaries, meaning that concentrations along the plane at $x=0$ equal those along $x = L_X$ (and similarly for $y = 0$ with $y = L_Y$ in a 3D domain).

However, as one of the goals of multi-d biofilm models is to rigorously compute and study the effect of 2d or 3d biofilm geometry on the external mass transfer rates, a mass-transfer boundary layer can be taken into account. On top of this boundary layer, at $z = L_Z$, the constant concentration $S = S_B$ is usually specified. If one assumes that only diffusion takes place in the boundary layer, the shape of the boundary layer can be assumed planar and parallel with the support surface (Figure 3.14B1 and also Figure 2.2B - this is the case of models N2a and N3a). Another possibility is to assume that the concentration boundary layer follows the biofilm surface in an attempt to mimic hydrodynamic effects (Figure 3.14B2 and Figure 2.2C - this is the case of model N3b with a zero boundary layer thickness, at the limit being the situation in Figure 3.14A).

If one chooses to compute rigorously mass transfer and hydrodynamics in the boundary layer by (3.108) and (3.111), then a set of boundary conditions shown in Figure 3.14C1 can be applied. Furthermore, if one decouples equations (3.108) in the liquid from (3.110) in the biofilm and solves the substrate balances separately in the two regions, then at the interface between the solid biomass region and the liquid region the equations are coupled by internal boundary conditions. Under the assumption that no interface reactions take place, the continuity conditions of the substrate concentration and of the diffusive flux at the biofilm surface Γ are: $S_L = S_F = S_{LF}$ (i.e., $S_{solid\ region} = S_{liquid\ region}$) and $(D\partial S/\partial n)_L = (D\partial S/\partial n)_F$, where $\partial/\partial n$ denotes the spatial derivative normal to the interface (Figure 3.14C2). This approach is used by Eberl *et al.* (2000a) and in model N3c. In some modeling studies (Noguera *et al.* 1999b), the influence of the hydrodynamics on the mass transfer is simplified by the assumption that the flow field induces a concentration boundary layer of constant thickness around the biofilm architecture (Figure 3.14B2). Consequently, advection and hydrodynamics are not explicitly taken into account, but implicitly in terms of an additional parameter, the concentration boundary layer thickness. Since the bulk region above this concentration boundary layer is assumed to be completely mixed, this not only is a work-around for discretization problems associated with diffusion-convection operators, but it also reduces the numerical effort, since the simulation of the bulk region (hydrodynamics and mass transfer) needs not to be carried out.

3.6.3.2.3 Momentum and mass balances for the liquid flow

The complication when using models including advective mass transport, $\mathbf{u}\nabla S$, is that the fluid velocity field $\mathbf{u}$ must either be known or calculated, and this usually is a computationally intensive task. The computational load is especially significant with model N3c, used for BM2. In the liquid region, the carrier of the advective transport is the hydrodynamic flow field, governed by the incompressible Navier-Stokes equations describing conservation of momentum and total mass of fluid

$$\frac{\partial \mathbf{u}}{\partial t} + \mathbf{u}\nabla\mathbf{u} = -\frac{1}{\rho}\nabla p + \nu\nabla^2\mathbf{u} + \mathbf{g}, \quad \nabla\mathbf{u} = 0 \qquad (3.111)$$

where $\mathbf{u}$ denotes the flow velocity vector (LT^{-1}), p is the pressure ($ML^{-1}T^{-2}$), ρ and ν denote constant density and kinematic viscosity, respectively, and $\mathbf{g}$ is the acceleration (MT^{-2}) caused by a body force (such as gravity). At the liquid-biofilm interface, typically no-slip boundary conditions are formulated for the flow field ($\mathbf{u}=0$). Mass balances in the form of equation (3.108) can be applied both in bulk liquid and in the biofilm region. Examples of such multi-d biofilm models including advection, diffusion, and reaction can be found in Dillon *et al.* (1995), Picioreanu *et al.* (2000a, 2000b, 2001), Dupin *et al.* (2001) and Eberl *et al.* (2001).

3.6.3.2.4 Mass balances for particulate components

As for the dissolved components, mass balances for particulate components (e.g., biomass) follow the same general balance equation (3.3). The net reaction terms r for each biomass type are computed in the same way as for dissolved substances, and the same stoichiometry matrix is used to represent the biomass transformation processes too. However, the biofilm models with discrete biomass representation used in this report use empirical approaches to the biomass transport term $\mathbf{j}$ in (3.3). As in 1d models, biomass transport can be caused by biomass *growth* and *decay*, biomass *division*, and *EPS production*. Furthermore, other processes changing the biofilm volume, such as biomass *detachment* and *attachment*, can also be considered in multi-d biofilm models.

1. Biomass growth and decay. It is assumed that the biofilm consists of n_X types of active biomass (numbered $i = 1,2,...,n_X$) and only one type of inert biomass (with $i = 0$ and denoted X_I) that results from the decay of the active biomass. The separation of active biomass into different categories is usually based on differences in metabolism. For example, in the case of BM3, aerobic heterotrophs X_H and aerobic autotrophic nitrifiers X_N co-exist. Grid-based models (e.g., N3a and N3b) treat biomass growth differently than do particle-based models (e.g., N2a).

Grid-based models (N3a, N3b). The space is divided in a grid of elementary volumes (most commonly squares in 2d and cubes in 3d) and the biomass concentration X is taken as state variable. The rate of biomass accumulation in each grid volume is then simply

$$\frac{dX_i}{dt} = r_{X,i}(\mathbf{S}, \mathbf{X}) \tag{3.112}$$

Particle-based models (N2a). The 3d biofilm structure is represented by a collection of n_P non-overlapping hard spheres of biomass, also called biomass particles. Each spherical particle p contains one type of active biomass and different fractions of other biomass components (e.g., inert biomass, storage polymers, etc.). The total mass $m_{X,i,p}$ (M/particle) of the particle p is, therefore, the sum of all biomass components, $m_{X,i,p}$. It is assumed that the density of the biomass type i in a biomass particle is $\rho_{X,i}$ (g biomass $i \cdot$ m^{-3} particle). When the biomass of the particle changes in time, volume and radius change accordingly. In the 2d model, the biomass particles are cylinders spanning the whole domain, i.e., length Δx and radius. This reads:

$$R_p = \left(\frac{3}{4\pi} \sum_{i=1}^{n_X} \frac{m_{X,i,p}}{\rho_{X,i,p}} \right)^{\frac{1}{3}} \text{ in 3d}, \quad R_p = \left(\frac{1}{4\pi\Delta x} \sum_{i=1}^{n_X} \frac{m_{X,i,p}}{\rho_{X,i,p}} \right)^{\frac{1}{2}} \text{ in 2d} \tag{3.113}$$

The growth of each biomass type i (active or inert) with time is described by an ordinary differential equation representing the mass balance for each biomass particle p:

$$\frac{dm_{X,i,p}}{dt} = r_{X,i}(\mathbf{S}, m_{X,i}) \quad \text{for each } p = 1,2,...,n_P \text{ and } i = 0,1,...,n_X \tag{3.114}$$

The net reaction rates for generation of biomass components, $r_{X,i}$, are typically functions of the active biomass of the particle, $m_{X,i,p}$, and concentrations $S_i(x,y,z)$ of various substrates present at the center (x,y,z) of the biomass sphere. An initial distribution of biomass particles at time $t = 0$ must be defined. It is assumed that all $n_{P,0}$ initial biomass particles have mass m_0 and, correspondingly, radius R_0. They are randomly distributed on the planar substratum surface with the centers at $z = R_0$.

2. Biomass division. Depending on the nutrients available in the environment and the kinetics of growth and decay processes, the bacterial mass contained in each grid volume (models N3a, N3b) or particle (model N2a) increases or decreases according to equations (3.112) or (3.114), respectively.

Particle-based models (N2a). The total biomass (i.e., active + inert in model N2a) in a spherical biomass particle is assumed limited to a maximum value, $m_X < m_{X,max}$ (g biomass · m^{-3} particle). $m_{X,max}$ is conveniently chosen to achieve the desired total biomass density in the biofilm X_F (g biomass · m^{-3} biofilm). When this maximum biomass $m_{X,max}$ in a sphere is reached, a new "daughter" sphere is created, touching the "mother" sphere in a randomly chosen direction. Part of the biomass contained in the "mother" is redistributed to the "daughter" sphere.

Grid-based models (N3a, N3b). A similar biomass division is performed also by the CA models (type N3a, N3b) when the biomass density X_i in a certain grid volume exceeds a critical value.

3. Biomass spreading. The way biomass transport mechanisms (i.e., biomass spreading) are implemented after division constitutes the main difference between grid-based and particle-based models.

Grid-based models (N3a, N3b). After biomass accumulation in a time step Δt, some amount of biomass from a grid element may be redistributed (transported) to other grid elements. The easiest way to model biomass transport (spreading) is by implementing a discrete dynamical model, often called cellular automaton (CA). Several CA algorithms for biomass spreading in a biofilm have been proposed in recent years. The N3a and N3b models use the algorithm described in Picioreanu *et al.* (1998a, 1998b).

Particle-based models (N2a). The biofilm spreading as a result of biomass growth occurs by shoving the spheres when they get too close to each other. In this model, the pressure build-up due to biomass increase is relaxed by minimizing the overlap of cells. The algorithms and methods used to shove the biomass particles to a mechanical equilibrium state are described in detail in Kreft *et al.* (1998, 2001) and Picioreanu *et al.* (2004). The value of this modeling approach is strengthened by the relative ease with which a mechanism for production and spreading of extracellular polymeric substances (EPS) can be implemented (Kreft and Wimpenny 2001). Moreover, different bacterial growth morphologies can be implemented, for instance the growth of filamentous bacteria causing poorly settling sludge (Picioreanu and Van Loosdrecht 2003; Martins *et al.* 2004). Biomass spreading in this model does not result in undue dispersal of clonal clusters; the extent of such mixing may bear on competition and cooperation within and between species in a biofilm.

4. Biomass detachment. Detachment has been implemented in a few different ways.

Grid-based models (N3a, N3b). A quantitative approach to biofilm detachment was developed in Picioreanu *et al.* (2001), based on laws of mechanics. It is founded on the hypothesis that biofilm failure will occur when the equivalent stress or biofilm cohesion strength, σ_e is larger than the ultimate tensile stress before fracture, σ_t (see, for example, Hibbeler 1991; Benham *et al.* 1996). In terms of normal (σ_X and σ_Y) and shear (τ_{YX}) stresses, the failure criterion in two-dimensional state of strain reads

$$\sigma_X^2 - \sigma_X \sigma_Y + \sigma_Y^2 + 3\tau_{YX}^2 \ = \ \sigma_e^2 \ > \ \sigma_t^2 \qquad (3.115)$$

The mechanical stress in the biofilm builds up due to forces acting on the biofilm surface as a result of liquid flow. For stress calculation in structures, the governing equations for mechanical equilibrium and compatibility are (Benham *et al.* 1996)

$$\frac{\partial \sigma_X}{\partial x} + \frac{\partial \tau_{YX}}{\partial y} = 0 \; ; \quad \frac{\partial \sigma_Y}{\partial y} + \frac{\partial \tau_{YX}}{\partial x} = 0 \; ; \quad \left(\frac{\partial^2}{\partial x^2} + \frac{\partial^2}{\partial y^2} \right)(\sigma_X + \sigma_Y) = 0 \qquad (3.116)$$

must satisfy the boundary conditions of the plane strain problem as defined by the applied liquid forces. These equations are solved to yield first the components of the strain tensor (ε_X, ε_Y, γ_{YX}) and then the stresses (σ_X, σ_Y, τ_{YX}) at required points in the structure. As boundary conditions, the normal and shear stresses on the biofilm surface must be specified. These can be calculated from fluid dynamics equations (3.111). Zero-displacement condition is set on the biofilm-carrier interface.

This model requires knowledge of mechanical properties of biofilms only scarcely measured until now, such as tensile strength (Ohashi and Harada 1994, 1996; Ohashi et al. 1999) and elasticity modulus (Stoodley et al. 1999). The two known biofilm detachment mechanisms, *erosion* (loss of small biofilm parts - eventually only cells - mainly from the biofilm surface) and *sloughing* (loss of massive biofilm chunks, often broken from the substratum surface), can be modeled by considering a single breakage criterion.

<u>Particle-based models (N2a)</u>

(a) The most simplistic way, with the sole purpose of keeping the biofilm thickness constant, consists in removing every particle that is shifted above an imposed biofilm thickness limit (i.e., when particle's center $z > L_F$) due to a shoving step. The removed particle is not relocated, but lost.

(b) A more general detachment mechanism involves modeling of mechanical stresses effective on the entire surface of the biofilm, such as erosion caused by shear forces, by using a continuous detachment speed function F_{de}. This method, which is a multidimensional extension of the method used in the 1D model N1, is described in Xavier et al. (2005a, 2005b). The value of the detachment speed function F_{de} in a point $\mathbf{x}=(x,y,z)$ located on the biofilm surface is defined by the equation:

$$u_{de} = \frac{d\mathbf{x}}{dt} = -F_{de}(\mathbf{x})\mathbf{n}(\mathbf{x}) \qquad (3.117)$$

where $\mathbf{n}(\mathbf{x})$ is the vector normal to the biofilm surface at point $\mathbf{x}$ and u_{de} is the global detachment velocity (LT^{-1}). Like in model N1, the velocity u_{de} yields a phenomenological description of the decrease of the biofilm thickness per unit time as result of *erosion*. The method used is flexible to allow F_{de} to take several forms. Local values of $F_{de}(\mathbf{x})$ can be dependent on any state variable such as the biofilm thickness, the surface shear stress, the local biofilm density or the local concentration of detachment-inducing chemical species. This method allows also discrete detachment events that derive from random instabilities in the biofilm surface, i.e. biomass *sloughing*, to be implicitly derived from the simulations. Sloughing is here implemented by removing patches of biofilm that lost the connectivity with the biofilm attached to the substratum.

3.6.3.3 Solution methods

For a long time, models capable to describing the spatially multi-d dynamics of mixed-species biofilms was hindered by their high computational demands. We owe the present rise of multi-d biofilm models not only to the more powerful computers, but also to newly developed, highly efficient numerical methods and computational algorithms.

3.6.3.3.1 The overall solution approach

One major problem is how to accommodate all the fast and slow physical, chemical, and biological processes. The answer comes from a time-scale analysis (Kissel et al. 1984), which shows that processes changing the biofilm volume (biomass growth, decay, and detachment) normally are much slower than processes involved in substrate mass balances (diffusion, advection, and reaction) (Picioreanu *et al.* 1999, 2000b). Therefore, the uncoupling of growth and reaction equations according to characteristic times is quite common in biofilm modeling and can be found in the early 1d biofilm models (Kissel *et al.* 1984; Rittmann and Brunner 1984). In addition, momentum transport (by advection or viscous dissipation) is much faster than the slowest step (i.e., diffusion) of substrate mass transfer.

Time-scale analysis justifies working with three time scales: (1) biomass growth, in the order of hours or days, (2) mass transport of solutes, in the order of minutes, and (3) hydrodynamic processes, in the order of seconds (see also section 2.6.3). In other words, while solving the mass balance equation for substrate, the flow pattern can be considered at pseudo-equilibrium for a given biofilm shape; at the same time, biomass growth, decay, and detachment are in frozen state. By exploiting the natural time-scale separation in biofilms, the step by which the whole algorithm advances in time is the one necessary for the slowest process, here biomass growth. Therefore, it is often sufficient to consider only the steady-state versions of the transport equations (3.108) and (3.111), i.e., the time-derivatives on the left-hand-sides are neglected. Only biomass governing equations like (3.112) or (3.114) need to be solved by time-stepping algorithms.

Due to time separation, the model processes can be executed in a sequential algorithm, as explained here:

- Set initial condition for particulate (biomass) components in the biofilm on the substratum
- Set initial condition for dissolved (substrates) in bulk liquid and biofilm
repeat
    - Solve hydrodynamics
        1. Momentum and mass balances for liquid flow [Eq. (3.111)]
    - Solve dynamics of dissolved components
        2. Mass balances of solutes in bulk liquid [Eq. (3.102)]
        3. Mass balances of solutes in biofilm [Eqs. (3.108)-(3.110) solved for steady state]
    - Solve dynamics of particulate components
        4. Biomass growth and decay [Eq. (3.112) or (3.114)]
        5. Biomass division
        6. Biomass spreading
        7. Biomass detachment [Eq. (3.116) or (3.117)]
    - Time step, $t \leftarrow t+\Delta t$
until $t > t_{end}$

3.6.3.3.2 Solution methods for particular parts of the model

Diffusion-reaction equations. By dropping the time-dependent accumulation term, the mass balances equations (3.109) and (3.110) become elliptic partial differential equations (PDE). After discretization with a finite-difference scheme in each point of the computational domain, the system of PDEs can be solved efficiently by a *non-linear multi-grid* method (Press *et al.* 1997). The multi-grid algorithm has the advantage of being simple, flexible, and very efficient for systems of elliptic PDEs, which usually occur in steady state problems. It is applied in models N3a, N3b, and N2a, based on the multi-species-multi-substrate biofilm model described in (Picioreanu et al., 2004; Xavier *et al.* 2005a, 2005b) and modified versions of the algorithms described in (Picioreanu *et al.* 1998a, 1998b).

Advection-diffusion-reaction equations. In models N2b and N3c, advective first-order terms appear in solute mass balances (3.108). Since there seems to be no unconditionally stable, positivity preserving, local, linear finite difference discretisation of second order or higher, one has to either use a first order upwind method (N3c) or a nonlocal High Order Compact (HOC) scheme. After discretization, the resulting system of non-linear equations can be solved by a Newton-Raphson procedure (Press *et al.* 1997). The linear system involved in the Newton-Raphson routine can be solved by a multi-grid technique or by a preconditioned Krylov subspace method (conjugated gradients - BiCG, BiCGSTAB, GMRES, etc.), an iterative method very adequate for large systems of linear equations with sparse matrix of coefficients. This solution method was applied in model N3c and in biofilm models by Eberl *et al.* (2000a). Alternatively, in model N2b the advection-diffusion-reaction equation was solved with a Lattice Boltzmann method (see next paragraph).

Hydrodynamic equations. In model N2b, momentum and mass-transfer equations are solved with lattice Boltzmann (LB) methods (Ponce Dawson *et al.* 1993; Chen and Doolen 1998). The main advantages of LB methods are the inherent algorithmic parallelism, programming simplicity, and physical soundness. It is relatively easy to incorporate irregular boundaries, and mesh generation is trivial, at least for the uniform grid LB. Disadvantages include the time-stepping procedure involved and rather empirical settings for boundary conditions. LB methods were applied in the advection-diffusion-reaction and hydrodynamic models of biofilms as described in Picioreanu *et al.* (2000a, 2000b, 2001) and Eberl *et al.* (2000b).

Model N3c uses for flow field calculations the classical artificial compressibility method of Chorin (Chorin 1967, Peyret and Taylor 1990) and a simple Finite Volume discretisation on a voxel grid. The diffusion-reaction equation (3.12) is discretized in the solid biofilm region by the standard central difference scheme (CDS) on the compact 7-point stencil. For the advection-diffusion equation (3.9), a simple first order upwind scheme is used for convective contributions and the second order cell centered scheme for the diffusion operator. While this scheme is only of low convergence order, it is optimal in some sense among linear difference approximations (Hundsdorfer and Verweer 2003).

Detachment equations. In models N2b, N3a, and N3b, detachment is caused by the mechanical stresses and strains generated by fluid forces acting on the biofilm. A finite-element method is used to solve the plane strain problem and to find the principal stresses in the biofilm structure. The type of finite element used is the parabolic isoparametric element. The basis of the method, a general algorithm and program code, is described in a textbook by Hinton and Owen (1977).

In model N2a, equation (3.117) is solved numerically by a method adapted from the procedure for monotonically advancing fronts introduced by Sethian (1996). A very efficient implementation of this detachment procedure in a biofilm model by using the fast marching level set method (Sethian 1996, 1999) was described in Xavier et al. (2005a, 2005b).

3.6.3.4 Software implementation

Although some models including hydrodynamic calculations were originally implemented for parallel computers (Picioreanu 1999; Picioreanu *et al.* 2000a; Eberl *et al.* 2000a, 2000b), all the models (except N3c) can now be efficiently run on ordinary personal computers. The computer codes are written usually in C/C++ or Fortran, but newer codes are also in Java and Matlab. One reality is that all the model codes were developed only for pure academic research. Therefore, their availability and user-friendliness are still restricted. On the other hand, a Java-based fully working version of model N2a for multiple dissolved and particulate components, implemented as described in Xavier et al. (2005a), can be freely downloaded from http://www.biofilms.bt.tudelft.nl/frameworkMaterial/monospecies2d.html. The same web site contains an interactive demonstration of this model.

3.6.4 2d models with discrete biomass and discrete solutes (the Cellular Automata models N2c, N2d, N2e, N2f)

A 2d, cellular automata (CA) model is used in the solution of the three benchmark problems. The CA biofilm model (Pizarro *et al.* 2001; Noguera *et al.* 2004) conceptually uses the same microscopic and macroscopic mass balances described at the beginning of this chapter, but differs from other models on the mathematical implementation of the problem. Rather than using differential equations to describe substrates and biomass gradients within the biofilm system, the CA model uses a completely discrete implementation, in which substrate and biomass are represented by individual particles of food and cells, which move around and interact in the simulated 2D domain according to local stochastic rules that represent diffusion, substrate utilization, microbial growth and decay, and convective flux.

Different from other CA biofilm models described in the literature (Wimpenny and Colasanti 1997; Hermanowicz 1998), the 2d CA model used in the benchmark analyses is fully quantitative, a property that is achieved by defining the masses of food and microbial particles and the stochastic rules that regulate their dynamic behavior in a way that is consistent with the fundamental mass balances that describe biofilm systems. The base model N2c includes substrate diffusion, microbial growth and decay, from which the more complex models N2d and N2e including advective transport and N2f for multispecies biofilms are derived.

3.6.4.1 Discretization of the physical domain

For the CA model to capture the physical characteristics of a biofilm problem, it is necessary to properly select the discretization variables (Δx and Δt) and the probabilities associated with the random walk of substrate particles, which represent molecular diffusion. In the CA model implementation, random walks of substrate particles mean that, at every time step, a food particle can change the direction of movement according to a simple stochastic rule: A particle has a probability p_0 of not changing direction, a probability p_1 of taking a 90 degree turn, a probability p_2 of reversing the direction of movement, and a probability p_1 of taking a 270 degree turn. These probabilities are restricted by the condition that $p_0+2p_1+p_2 = 1$ (Pizarro *et al.* 2001; Noguera *et al.* 2004). In the simplest case of a single substrate diffusing on a homogeneous domain, $p_0=p_1=p_2=0.25$.

The key connection between the probabilities that define the random walk of substrate particles, the diffusion coefficient (D), and the discretization of a 2d domain in space and time is given by Equation (3.118) (Chopard and Droz 1991). For a given lattice grid size (Δx) and specific probability values p_0 and p_1, Equation (3.118) defines a unique time step Δt to ensure that the random walk of food particles is physically representing the diffusion of a substrate characterized by having a diffusion coefficient D.

$$D = \frac{\Delta x^2}{\Delta t} \frac{p_0 + p_1}{4\left(1 - \left(p_0 + p_1\right)\right)} \tag{3.118}$$

When multiple substrates, characterized by having different diffusion coefficients, are diffusing within the same domain, equation (3.118) should be valid for all substrates. Once Δx and Δt have been selected, it is possible to select different p_0 and p_1 values for the different substrates, as described in Noguera *et al.* (2004).

3.6.4.2 Definition of substrate and microbial particles

Once the size Δx of an element within the 2D domain has been determined, then the total volume of a CA element is defined by $(\Delta x)^3$. The mass of substrate contained within a substrate particle is then given by equation (3.119), while the mass of biomass contained within a microbial particle is represented by equation (3.120).

$$m_S = \frac{S_B}{N_m}(\Delta x)^3 \tag{3.119}$$

$$m_X = X_{F,max}(\Delta x)^3 \tag{3.120}$$

In these equations, S_B is the concentration of substrate in the bulk liquid, N_m represents the number of layers within the substrate lattice, (i.e., $N_m=4$ in 2D, cf. Noguera *et al.* (2004)), and $X_{F,max}$ is the maximum possible concentration of biomass within the biofilm system. The CA biofilm model simulates random walks of substrate particles using a technique called partitioning, which was developed by D'Souza and Margolus (1999). In this technique, it is necessary to allow the potential for several particles to be co-localized in the same CA element at a given time. Accordingly, in a 2d domain, the CA elements are divided into four layers, and hence $N_m = 4$. Details for the computational implementation of the partitioning technique can be found elsewhere (Pizarro *et al.* 2001, 2005; Noguera *et al.* 2004).

3.6.4.3 Discretization of Monod-type substrate-utilization kinetics

In the CA model, substrate utilization kinetics is represented by a stochastic rule that determines the probability of a substrate particle being consumed during one time step. This probability, p_u, is calculated according to equations (3.121), which define a discretized version of a Monod-type kinetic expression (Pizarro *et al.* 2001; Noguera *et al.* 2004). In this equation, μ, μ_{max}, K, and Y represents the typical kinetic parameters used to describe microbial growth and substrate utilization according to a Monod-type expression (defined in Section 2.4.1), S_f is the local concentration on a particular node of the 2D grid, and is defined by equation (3.122) as the ratio of the number of substrate particles present the node (N_p) over the maximum number of substrate particles that can be present in that node (N_m), times the bulk substrate concentration (S_b). Pizarro *et al.* (2001) demonstrated that this type of stochastic approximation adequately represents Monod-type kinetics for substrate concentrations as high as ten times K.

$$p_u = \frac{\mu}{Y}\frac{(\Delta x)^3 \Delta t}{m_S} = \mu_{max}\frac{S_F}{K+S_F}X_{max}\frac{(\Delta x)^3 \Delta t}{m_S} \tag{3.121}$$

$$S_F = S_B\frac{N_p}{N_m} \tag{3.122}$$

3.6.4.4 Stochastic representation of microbial growth, inactivation, and endogenous respiration

In the CA biofilm model, microbial growth is stochastically represented as the probability p_g of a microbial particle to duplicate during a time step. This probability is proportional to the probability of substrate utilization p_u, according to equation (3.123) (Noguera *et al.* 2004). The probability of inactivation is estimated from the rate laws according to equation (3.124), while the probability of microbial particles disappearing because of endogenous respiration is calculated using the corresponding rate laws, as illustrated in equation (3.125).

$$p_g = \left(\frac{Y m_S}{m_X} \right) p_u \tag{3.123}$$

$$p_{ina} = \frac{r_{ina}}{m_X} (\Delta x)^3 \Delta t \tag{3.124}$$

$$p_{res} = \frac{r_{res}}{m_X} (\Delta x)^3 \Delta t \tag{3.125}$$

3.6.4.5 Simulation of microbial dynamics within the biofilm (N2f)

Different rules can be implemented for the distribution of microbial particles within the biofilm, as described in detail by Noguera *et al.* (2004). For the benchmark problems described in this report, BM3 is the only case where it is necessary to simulate microbial dynamics within the biofilm. To optimize computational efficiency of the model, the microbial lattice is updated with a larger time step ΔT, thus taking advantage of the characteristic times concept described in Section 2.6.3. For this large time step, the probabilities of growth, inactivation, or respiration events occurring during the large time step are adjusted by multiplying the probabilities calculated from equations (3.123)-(3.125) by the factor $\Delta T / \Delta t$ (Pizarro *et al.*, 2001).

In BM3, three different types of microbial particles need to be considered: heterotrophs, autotrophs, and inactive cells. In the CA model, one microbial lattice stores the location of these microbial particles. When a microbial particle reproduces, the new particle is stored in an auxiliary microbial particle lattice. Then, the distribution of newly created microbial particles is performed assuming a random walk of newborn particles, in a similar way as how the random walk of substrate particles is implemented, using the partitioning technique mentioned above, and described in detail in Noguera *et al.* (2004). During this random walk, special care needs to be taken to ensure that newborn particles stay near the location where they are born. Accordingly, after each step in the random walk movement, newborn particles are placed on the main lattice, if there is available space. If the main lattice is occupied, then a swapping of a newborn particle with the microbial particle present in the main lattice is implemented, and the random walk process is continued. This procedure allows newborn particles to remain close to were they were created, simulation an effect of "pushing" cells within the biofilm.

A final step in the redistribution of microbial particles is implemented in BM3 to maintain the density of the biofilm constant. If spaces are left within the biofilm (from endogenous respiration) after the redistribution of newborn particles, they are eliminated by collapsing the biofilm and closing the gap.

3.6.4.6 Simulation of advective flux (N2d, N2e)

For the solution of BM2, advective flux is simulated with a discretized approximation compatible with the CA model. In this case, it is assumed that the fluid velocity throughout the 2d domain can be represented by a linear gradient, with u_0 being the velocity at the top boundary of the domain (as defined in the BM2 problem statement), and having a zero velocity at the surface of the base biofilm. To convert this fluid flow component to local rules for the CA model, it is necessary to calculate how many discrete substrate particles (or "locations" of substrate particles, if the lattice element was empty) should be moved at each time step.

To represent the linear velocity gradient, the number of food particle locations to be displaced at each time step depends on the distance of the element to the top of the domain.

Finally, to eliminate errors caused by rounding the displacements to the nearest integer, especial attention is taken to assure that the total number of displacements represents the integrated advective flux into the domain. When necessary, small corrections to the rounding process are made.

3.6.5 Applications

Just as the N1 model, the N2 and N3 multi-d models are flexible frameworks for description of many types of biofilm systems (Xavier *et al.* 2005a). The framework approach provides a tool in which any type of microbe/food kinetic interaction may be implemented, together with any number of solute and particulate species involved. The biomass composition in the models created using this framework may be as detailed as necessary, including, for example, active biomass, inerts, and EPS. Like 1d models, the N2/N3 models can also describe the dynamics of solute and particulate components and, therefore, substrate removal and biomass production rates in a dynamics biofilm system. In addition, however, the N2/N3 models are particularly suitable for the study of new topics, such as: (i) biofilm structure formation in relation with environmental factors, (ii) evaluation of diffusive and convective substrate transport as a result of fluid flow, (iii) 3d spatial interactions between microbial communities, or (iv) the role of bacterial motility in biofilm formation.

3.6.5.1 Formation of biofilm structure and activity in relation with the environment

3.6.5.1.1 Biomass accumulation in biofilm

The total biomass accumulation in biofilm is an important indicator of biofilm structure and is calculated for the N2a, N3a, and N3b models by summation of all particulate biomasses over a certain substratum area. If the biofilm thickness is not leveled off by detachment, the multi-d models predict less biomass than 1d models. Reduced biomass formation can be seen in Figure 3.15, where the total biomass is shown only until the biofilm reaches 500 μm. For "narrow" 2d systems, where the length L_X is so small that the surface instability cannot develop, a pseudo-1d system is obtained as a degenerated case of the multi-d system. The total biomass and bulk concentrations obtained with such a pseudo-1d model (stars in Figure 3.15) are approximately the same as those calculated with the 1d model N1 (lines in Figure 3.15). As discussed in Picioreanu *et al.* (1998a, 2004), 2d and 3d models based only on the three processes of diffusion, reaction, and growth suggest that formation of porous biofilm structures is favored under nutrient-limitation (see also Figure 2.5 and Eberl *et al.* 2001). This can be explained as a consequence of the strong substrate gradients, which lead to much greater substrate fluxes (Figure 3.16) and consequently greater microbial growth rates at the top than at the bottom of the biofilm (see also Figure 3.17). When the initial microbial distribution on the surface is non-uniform, a wavy biofilm surface will develop due to a self-enhancing process. Intrinsically, 1d biofilm models (such as N1) assume a planar biofilm surface and concentration boundary layers, as well as 1d substrate gradients and fluxes. Therefore, an unstable biofilm surface is impossible, and a compact layer with a higher total biomass per area will develop due to the absence of substrate-depleted valleys. More discussions on the results generated by N3a and N3b models and of N2a model can be found in Chapter 4, benchmark problems BM1 and BM3, respectively.

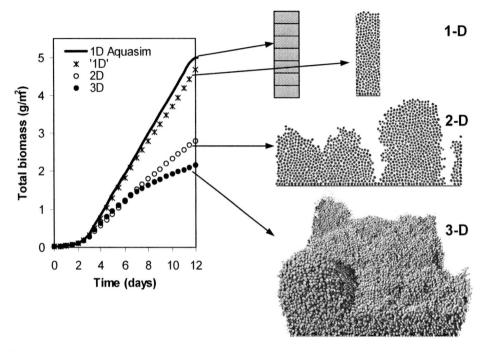

Figure 3.15. Comparison of total biomass accumulated in the biofilm ($X_{F,A}$) calculated with a 1D model (N1) and with 2D/3D models (N2a, N3a, N3b) without biofilm detachment (data from Picioreanu *et al.* 2004).

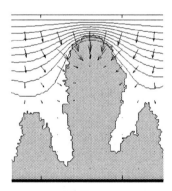

Figure 3.16. Substrate fluxes (arrows) to a biofilm (gray area) with irregular surface shape (from Picioreanu *et al.* 1998a).

3.6.5.1.2 Substrate uptake and mass transport in biofilm systems

The contribution of pores and channels to the transport to the bacterial cells can be assessed by simulating the substrate conversion process at different flow rates and biofilm geometries. Such an evaluation is made in BM2 (Section 4.3), for multi-d biofilm models N2b, N2d, N2e, and N3c. One requires knowledge of liquid flow pattern past the biofilm (see Figure 3.17), which affects the flux of substrate transport by convection. Due to the inherent computational complexity, accurate hydrodynamics has rarely been considered in biofilm modeling (e.g., Dillon *et al.* 1996; Dillon and Fauci 2000; Picioreanu *et al.* 1999, 2000a, 2000b, 2001; Eberl *et al.* 2000a, 2000b). The effect of flow velocity on the relative contribution of advective and diffusive transport mechanisms to the overall substrate transport to the same biofilm is shown in Figure 3.18 (Picioreanu *et al.* 2000a). Only at very high flow rates does advection dominate substrate transport. At such high flow rates, however, also large shear forces exist, and the biofilm will adapt by becoming less porous and smoother, decreasing the pore-based advective transport. The biofilm structure seems to adapt to the flow regime in a way that buffers mass transfer. This implies that advective transport inside a biofilm might not be very important in general, unless large changes in flow rates occur in a biofilm system or the biofilm structure contains large moving parts (Stoodley *et al.* 1998).

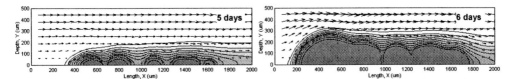

Figure 3.17. Simulated biofilm development in time under 2d hydrodynamic conditions. The arrows represent the field of fluid velocity; Contour lines like in Figure 3.18. (from Picioreanu *et al.* 2000b)

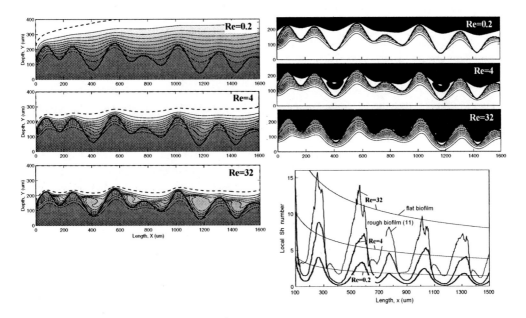

Figure 3.18. Effect of flow velocity on the relative contribution of convective and diffusive mass transport. Left: Contour lines of substrate concentration at different fluid velocities (different Re numbers) for a rough surface biofilm. The thick continuous lines indicate the biofilm surface. Iso-concentration lines show the decrease of substrate concentration from the maximum value in the bulk liquid (white areas) to zero in the biofilm (dark-grey areas). The thick dashed contour lines indicate the limit of the concentration boundary layer (98% from the bulk concentration). Right: Advection dominates mass transfer in black areas. Inside the biofilm structure lines of equal reaction rate are drawn. The graph shows the local flux of substrate through the biofilm surface. Model description and parameters are in Picioreanu *et al.* (2000a).

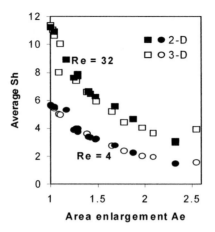

Figure 3.19. Effect of biofilm heterogeneity, expressed as area enlargement (biofilm surface area divided by substratum surface area), on mass transfer expressed as Sherwood number (total external mass transfer divided by diffusive transfer). Comparison between 2d and 3d simulations at relatively low (Re=4) and high (Re=32) flow velocity (data from Eberl *et al.* 2000a).

It can be argued that the above simulations by a 2d model are not representative of a 3d structure, where flow can by-pass dense biofilm structures. Simulations by Eberl *et al.* (2000a) show that, when the same biofilm is modeled in 2d or 3d, the overall mass transfer is equivalent (Figure 3.19). This indicates that, for many studies, a 2d simulation is sufficient. The effect of 3d geometry and porosity has been further evaluated by simulating a mushroom-like biofilm. Again, it becomes apparent that, in the pore region, at relatively high liquid velocities, mass transfer is dominated by advection. However, the exchange of liquid between the pore region and the bulk liquid is marginal. Therefore, if the overall mass transfer from the bulk liquid to the biofilm is evaluated, diffusive transport dominates advective transport even at high liquid velocities. An exception is a biofilm consisting of isolated colonies; advective transport then contributes significantly to mass transfer. In conclusion, these model results indicate that pores probably do not contribute much to the overall conversion process (where the interest is usually from an engineering perspective), although they might have local influences on microbial competition and selection processes (where the interest of microbiologists often is).

3.6.5.1.3 Biofilm structure

The geometrical structure of biofilms is determined, among many factors of biological or physical-chemical nature, by the balance between biomass growth as a result of nutrient availability and biomass detachment as a result of erosion and sloughing (Van Loosdrecht *et al.* 1995b). The effect of nutrient availability on development of compact or porous biofilms was described in Section 3.6.5.1.1. On the other hand, biofilm detachment accounts for biomass losses from processes acting continuously on the entire surface of the biofilm, such as erosion, and discrete detachment events, such as sloughing. Simulations performed with models N3a, N3b, and N2a show that high detachment rates or low biomass growth rates lead to the formation of a smooth biofilm structure (e.g., Figure 3.20A – 3d simulation; Figure 3.20C – 2d simulation), whereas low detachment rates or high biomass growth rates generate a rougher biofilm aspect (Figure 3.20B – 3d simulation; Figure 3.20D – 2d simulation).

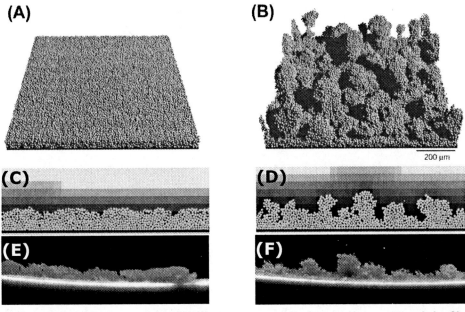

High detachment rate -> **Smooth** biofilm   **Low** detachment rate -> **Rough** biofilm

Figure 3.20. High detachment rates lead to the formation of a smooth biofilm structure (A – 3d simulation; C – 2d simulation; E – experimental), whereas low detachment rates generate a rougher biofilm aspect (B – 3d simulation; D – 2d simulation; F – experimental). 3d and 2d simulations performed as described in (Xavier *et al.* 2005b). A Java-based 2d biofilm simulator can be found at http://www.biofilms.bt.tudelft.nl/frameworkMaterial/monospecies2d.html. The real biofilm pictures are from nitrifying rotating drum reactors, cf. Kugaprasatham *et al.* (1992).

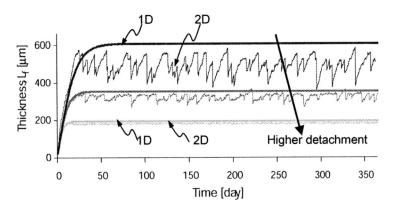

Figure 3.21. Time course of biofilm maximum thickness, L_F, resulted from simulations at different applied detachment rates with a 1d dynamic model (thick lines; AQUASIM, model N1) and a 2d biofilm model (thin lines; results from Xavier *et al.* 2005c).

Simulation results range from approximately constant L_f values for high detachment (erosion) rates to very noisy L_f values observed in the lower detachment regimes (Figure 3.21). The same biofilm thickness is calculated by the 1d and 2d models at high detachment forces, when the 2d model generates basically a planar biofilm (Figure 3.20). However, the two modeling approaches diverge for the cases where the applied detachment rates are low.

3.6.5.2 Model comparison with experimental data

In spite of obvious progresses in multidimensional modeling of biofilms, the necessity to compare model predictions with experimental data is becoming more critical (Kreft *et al.* 2001). This necessity was approached in a recent study (Xavier *et al.* 2004) were the performance of a 3-D model was assessed by comparison with biofilm structure observed experimentally from confocal laser scanning microscopy (CLSM) monitoring of a biofilm grown in a laboratory flowcell. CLSM imaging of biofilms is an ideal source of data for the evaluation of 3-D biofilm structure model predictions, due to its dynamic but non-destructive characteristics. However, due to the stochastic nature of biofilm development, evaluation cannot be carried out by direct comparison of the 3-D spatial structure. A quantitative comparison, therefore, imposes the use of morphological parameters such as those described previously in the literature which, thanks to the similar nature of the 3-D data obtained both from CLSM imaging and the model simulations, could be readily determined by procedures derived from automated image analysis operations (Xavier *et al.* 2003). The study by Xavier *et al.* (2004) concluded that 3d models are capable of accurate prediction of biofilm structure, as described by a set of morphological parameters such as filled space fraction, total biomass, substratum coverage, mean biofilm thickness and surface roughness (Figure 3.22).

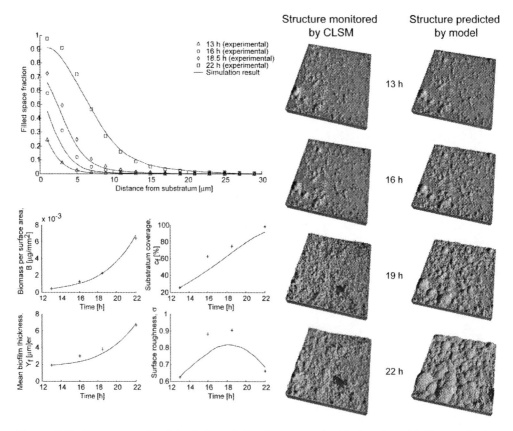

Figure 3.22. Structure predicted by 3-D modeling is compared with data from biofilm monitoring using CLSM (Xavier *et al.* 2004, 2005c). The stochastic process of biofilm development, which is also implemented in the model, requires the use of morphology parameters to perform the comparison. The initial biomass distribution for the simulated biofilm is taken from the experimental biofilm.

3.6.5.3 Interactions in multispecies biofilms

Studying interactions between diverse microbial types in a biofilm community is an important application of multi-d biofilm models. In this report, some of the results produced by models N2a and N2f are compared in benchmark problem BM3 (Chapter 4.4). Examples of such studies have appeared relatively recently and include biofilm systems with: methanogens (Noguera *et al.* 1999b), nitrification (Kreft *et al.* 2001; Picioreanu *et al.* 2004), methanogens in anaerobic digestion granules (Picioreanu *et al.* 2005a, 2005b), and competition between EPS- and PHB-producing organisms (Xavier *et al.* 2005a).

3.6.5.3.1 Multispecies nitrifying biofilm

A nitrifying biofilm involving three microbial species – aerobic ammonium oxidizers, aerobic nitrite oxidizers, and anaerobic ammonium oxidizers – was implemented in a particle-based variant of the individual-based modeling approach (Picioreanu *et al.* 2004). The system describes also the decay of active biomass to inert materials and includes five solute species – oxygen, ammonium, nitrite, nitrate, and N_2. Simulations were performed with different scenarios of availability of ammonium and oxygen in order to evaluate the evolution of the three microbial species present. Stratification of the biofilm according to the presence of electron acceptor and the growth rate of microorganisms was observed, showing different distributions for the different growth conditions (Figure 3.23). For example, long-term simulations suggest that at high dissolved oxygen concentrations (10 mg/L) all three groups can coexist in the biofilm (Figure 3.23A), but at lower dissolved oxygen (2 mg/L), the aerobic nitrite oxidizers may be completely eliminated from the biofilm by the faster growing anaerobic bacteria (Figure 3.23B).

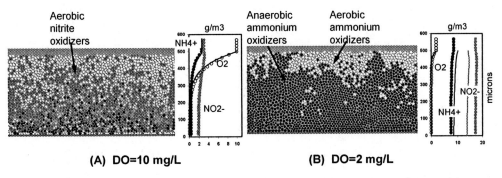

Figure 3.23. Simulation results produced by model N3e for a multi-species nitrifying biofilm after more than 200 days of biofilm development. The concentration of dissolved oxygen in the bulk was (A) 10 mg/L (biofilm mostly aerobic) and (B) 2 mg/L (biofilm largely anaerobic). In the 2d biomass distribution maps, there are aerobic ammonium oxidizers (white circles), aerobic nitrite oxidizers (gray circles) and anaerobic ammonium oxidizers (black circles). Vertical profiles show concentrations of dissolved species in the biofilm depth, with lines representing values computed with the 1d AQUASIM model N1 and symbols showing average values obtained by the 2d biofilm model (from Picioreanu *et al.* 2004).

Comparisons carried out in the same study between 1d (N1 model implemented in AQUASIM), 2d and 3d modeling (N2a model) alerted for the presence of strong multidirectional gradients likely to be found for the concentration of intermediate solutes. This was the case of nitrite, which, being produced by ammonium oxidizers and consumed by nitrite oxidizers, showed significant gradients in the directions parallel to the solid surface in 2d (Figure 2.4A) and 3d simulations (Figure 2.4B), a characteristic obviously not to be observed using 1d modeling. This comparative study, nevertheless, showed good agreement between all modeling approaches in respect to overall conversion rates. This indicates that 1d modeling approaches may be sufficient to accurately describe bioconversions occurring in a planar (flat surface) biofilm system. 2d and 3d approaches, however, provide further insight in respect to local microbial interactions of competition and cooperation occurring in the system. Other recent modeling studies arrived at the same conclusion for mono-species biofilms (Morgenroth *et al.* 2004) and for the heterotrophic-autotrophic biofilm system analyzed in BM3 (Section 4.4 and Noguera and Picioreanu 2004).

3.6.5.3.2 Multispecies methanogenic biofilm

Another problem with great potential for spatial modeling is the ecology of methanogenic communities (Picioreanu *et al.* 2005a, 2005b), such as in understanding the kinetic of anaerobic digestion and interpreting results obtained from advanced microbial ecology analysis (using molecular methods). Formation of a layered structure of the granules or biofilms has been proposed, based on the relative kinetic rates of the different steps in anaerobic digestion (AD). AD is a multi-step process, with (in order), extracellular hydrolysis, acidogenesis, acetogenesis, and hydrogenotrophic and aceticlastic methanogenesis (Rittmann and McCarty 2001). A good base to think about these interactions is the anaerobic digestion model proposed by an International Water Association (IWA) task group (Batstone *et al.* 2002). Acidogenic bacteria convert monosaccharides, amino acids, and long-chain fatty acids into short-chain fatty acids such as valeric, butyric, or propionic. Then, syntrophic organisms produce acetic acid and hydrogen, which are finally converted to methane. An example of such a simulation including seven bacterial groups and twelve chemical species is presented in Figure 3.24A. The growth of syntrophic acetogens (propionate and butyrate utilisers) is of particular interest. When growing near the biofilm surface (or the edge of the methanogenic granule), these microbes can live without the presence of hydrogenotrophic methanogens, as hydrogen can be wasted to the bulk. It should be noted that releasing hydrogen to the bulk only helps if hydrogen is continuously removed from the bulk, and this is the case assumed here. When growing in the biofilm depth, they only survive in the presence of hydrogen utilisers. This is explained by the fact that the syntrophic propionate consumers are inhibited by the hydrogen produced by themselves and by the sugar-converting acidogens. Therefore, they prefer to grow in the vicinity of the hydrogenotrophs, which in turn depend on the acidogens for hydrogen production. Multi-d fluxes of hydrogen in such a methanogenic biofilm can be seen in the simulation presented in Figure 3.24B.

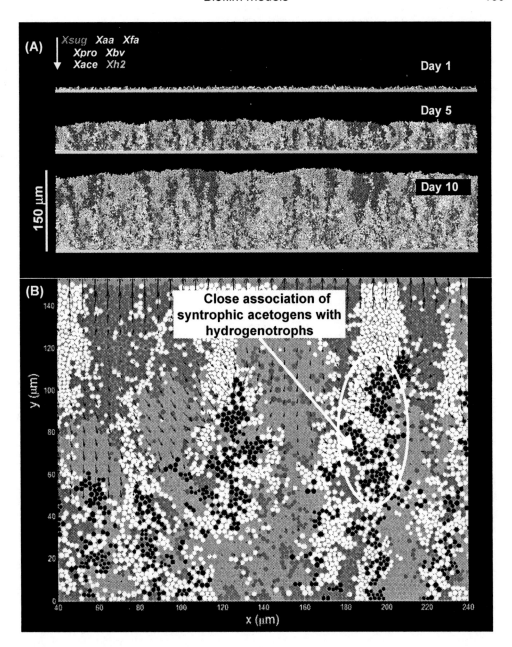

Figure 3.24. (A) Simulated development of an anaerobic digestion (methanogenic) biofilm. The seven microbial groups are: (1) acidogens (dark gray circles) growing on sugar (Xsug), amino acids (Xaa), fatty acids (Xfa), (2) syntrophic acetogens (white circles) growing on butyrate/valerate (Xbv), propionate (Xpro), (3) methanogens on acetate (light gray circles) (Xace), and (4) methanogens growing on hydrogen (Xh2, black circles). Formation of vertical bacterial clusters and horizontal stratification with fast growing microorganisms near the biofilm surface (sugar and propionate consumers) can be observed. (B) Close-up in a methanogenic biofilm shows intimate association of syntrophic acetogens - inhibited by H2 - (white) with hydrogenotrophic methanogens - consuming H2 - (black). Arrows indicate the direction and magnitude of hydrogen fluxes.

3.6.5.3.3 Competition between internal storage compound producer and EPS producing bacteria

The potential of the multi-d biofilm models in handling structured biomass, extracellular polymeric substances (EPS), and variable feeding conditions is illustrated in a case study describing the competition between two bacterial species differing in their ability to produce either PHB (polyhydroxybutyrate, an intracellular storage polymer that constitutes a reserve of carbon and energy) or EPS. A hypothetical system was proposed in Xavier *et al.* (2005a): two competing heterotrophs, PHB-producers and EPS-producers, possess the same kinetics for growth and production of a polymer, PHB or EPS, respectively (Figure 3.25). For this two-species system, growth in a chemostat should offer no competitive advantage to either organisms, as their kinetics are identical. In a biofilm in which growth is limited by diffusion of oxygen, EPS-producers have a competitive advantage, as shown by simulations. By spreading more rapidly due to the production of a voluminous EPS matrix, EPS-producing organisms take advantage of higher oxygen concentration at the top of the biofilm, eventually overcoming the PHB-producers.

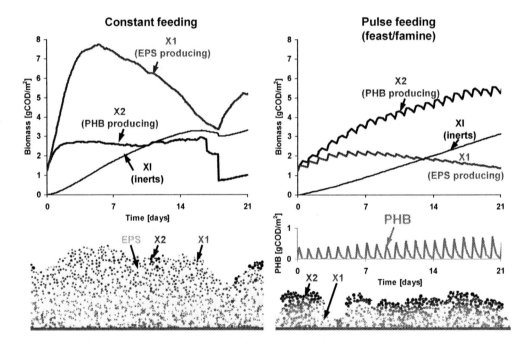

Figure 3.25. Results from 2d simulations of a two-species biofilm were the competition between PHB producing and EPS producing heterotrophic microorganisms is observed under different substrate feeding regimes (Xavier et al. 2005a). EPS-producers (X1 - dark gray) have a competitive advantage over PHB-producers (X2 - black) when carbon source feeding is constant, in spite of a similar kinetics for both species (EPS is shown in light gray). The competition is reversed by growing the community in a feast/famine regime: PHB-producing organisms accumulate PHB internal storage during the feast phase and use this storage during the famine phase when growth of EPS-producers is halted due to the absence of an external carbon source.

Additional simulations show that this competition can be inverted in feast-famine regimes, where the organic-donor substrate is provided in pulses. PHB-producers become the more competitive species thanks to their ability to produce the internal storage compound during the feast phase and latter using that stored PHB for growth during the famine phase. EPS-producers, not possessing the capability to produce storage compounds, only grow in the feast phase and starve and detach in the famine phase. Biomass detachment by continuous surface erosion is implemented using an approach explained in Xavier et al. (2005b). This system considers two solute species – oxygen and a soluble carbon source – and five particulate species – active mass of PHB-producers, PHB, active mass of EPS producers, EPS and inert biomass – in a structured representation of biomass.

4

Benchmark problems

4.1 INTRODUCTION

This chapter compares the performance of the biofilm models described in Chapter 3 for solving three characteristic benchmark problems. The multi-dimensional dynamic models describe more aspects of the behavior of biofilm systems than do simple steady state models. At the same time, the multi-dimensional models require more input data and lead to higher computational costs. The benchmark problems help identify the trade offs inherent to using the different types of models. Because each model solves the same benchmark problem, the differences of the output produced by the models reflect the differences of the model complexity and of the simplifying assumptions that are made in the various models.

The first benchmark problem (BM1, Section 4.2) describes nearly the simplest system possible: a mono-species biofilm that is flat and microbiologically homogeneous. Benchmark problem 2 (BM2, Section 4.3) evaluates the influence of hydrodynamics on substrate mass transfer and conversions in a geometrically heterogeneous biofilm. Benchmark problem 3 (BM3, Section 4.4) describes competition between different types of biomass in a multi-species and multi-substrate biofilm. Because the benchmark problems were designed to evaluate the ability of the models to represent fundamental features of a biofilm system, the trends apply to biofilms in treatment technology, nature, and situations in which biofilm is unwanted. The results obtained from the benchmark problems informed the Task Group as it prepared it guidance for model selection in Chapter 2.

4.2 BENCHMARK 1: SINGLE-SPECIES, FLAT BIOFILM

The first benchmark problem 1 (BM1) describes a simple flat, mono-species biofilm. BM1 gives a baseline comparison of the different biofilm models for a biofilm system that is well suited for any modeling approach. The specific objective of BM1 is to compare key outputs, particularly including effluent substrate concentrations and substrate flux. Furthermore, the user friendliness of the different modeling approaches is evaluated.

4.2.1 Definition of the system to be modeled

In BM1, the biofilm system has one particulate component (active heterotrophic biomass) and two soluble components (organic substrate and oxygen) as growth-limiting substrates. The biofilm accumulates on the walls of a rectangular channel, as illustrated in Figure 4.1. The parameters describing the physical system are summarized in Table 4.1.

Table 4.1. Reactor and physical biofilm parameters

Parameter	Symbol	Value and Units
Flow rate	Q	0.02 m³/d
Reactor length	L_R	0.5 m
Reactor width and height	H_R	0.05 m
Biofilm surface area	$A_F = 4H_R L_R$	0.1 m²
Reactor volume	$V_R = H_R^2 L_R$	1.25 x 10⁻³ m³
Biomass in the reactor	M_X	0.5 g_{COD-X}
Biomass density in the biofilm	X_H	1 x 10⁴ g_{COD-X}/m³
Biofilm volume	$V_F = M_X/X_F$	5 x 10⁻⁵ m³
Biofilm thickness	$L_F = V_F/A_F$	500 μm
Boundary layer thickness	L_L	0
Substrate (COD) influent concentration	$S_{in,S}$	30 g_{COD-S}/m³
Oxygen concentration in the bulk liquid	S_{O2}	10 g_{O2}/m³
Substrate diffusion coefficient in pure water	D_S	1 x 10⁻⁴ m²/d
Oxygen diffusion coefficient in pure water	D_{O2}	2 x 10⁻⁴ m²/d

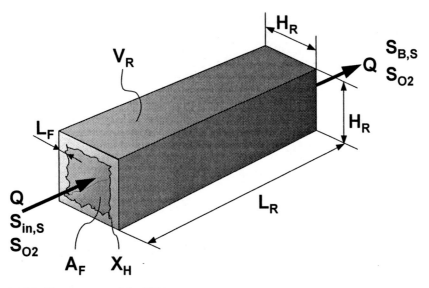

Figure 4.1. Biofilm reactor used for BM1.

The following assumptions underlie the <u>standard case</u> of BM1:

1. The biofilm reactor is a long rectangular channel completely filled with water flowing through it. A section of length $L_R = 0.5$ m of this channel is modeled.
2. The bulk liquid of the modeled section is completely mixed.
3. A heterotrophic biofilm accumulates on all flat channel walls. The total biomass in this reactor section is set to $M_X = 0.5$ g_{COD-X}, and the average biomass density in the biofilm is $X_F = X_H = 10000$ g_{COD-X}/m^3. Irregularity of biofilm geometry at the channel edges is neglected. An average biofilm thickness $L_F = 500$ µm results from the combination of total biomass, biofilm density, and biofilm area. Different modeling approaches consider the biofilm either flat with a constant thickness L_F or a heterogeneous structure with an average thickness equal to L_F.
4. Biomass inflow at concentration $X_{in,H}$ to the reactor section is negligible.
5. S denotes readily degradable organic substrate, and its concentration in inflow is $S_{in,S} = 30$ g_{COD-S}/m^3. Substrate concentration in the effluent $S_{ef,S}$ is equal to $S_{B,S}$ in the bulk.
6. The concentration of O_2 in the bulk liquid is kept constant by aeration at $S_{B,O2} = S_{O2} = 10$ g_{O2}/m^3.
7. The diffusion coefficients of the substrate and of dissolved oxygen in the biofilm are equal to the diffusion coefficients in pure water, respectively.
8. Mass-transfer resistance at the interface between the biofilm and bulk fluid are neglected ($L_L = 0$).
9. The biofilm system is at steady state, i.e., all variables and parameters are constant in time.
10. Heterotrophic biomass grows using an organic substrate as the electron donor and oxygen as the electron acceptor.
11. A decay process reduces the amount of active biomass.
12. The net biomass detachment rate equals the net biomass growth rate (i.e., the biofilm thickness is steady state). In other words, when the synthesis of active biomass exceeds decay for the entire biofilm, then the specified average biofilm thickness is maintained by detachment.
13. Because the biofilm has only one particulate component (heterotrophic active biomass), the biofilm is, by definition, microbiologically homogeneous.
14. No conversion processes take place in the bulk liquid.

Table 4.2 summarizes the stoichiometry and kinetics that correspond to these assumptions. The kinetic and stoichiometric parameters of the microbial processes are summarized in Table 4.3.

Table 4.2. Matrix of stoichiometric coefficients and rates of microbial processes

Process	Components			Process rate [$ML^{-3}T^{-1}$]
	X_H	S_S	S_{O2}	
Heterotrophic growth	1	$-\dfrac{1}{Y_H}$	$-\dfrac{1-Y_H}{Y_H}$	$\mu_{max,H} \dfrac{S_S}{K_S + S_S} \cdot \dfrac{S_{O2}}{K_{O2} + S_{O2}} X_H$
Heterotrophic decay	-1			$b_H X_H$

Table 4.3. Kinetic and stoichiometric parameters for heterotrophic biomass

Parameter	Symbol	Value and Units
Maximum specific utilization rate	$\mu_{max,H}$	6 d^{-1}
Decay rate coefficient	b_H	0.4 d^{-1}
Substrate half-maximum rate concentration	K_S	4 g_{COD-s}/m^3
Oxygen half-maximum rate concentration	K_{O2}	0.2 g_{O2}/m^3
True yield of biomass produced from substrate	Y_H	0.63 g_{COD-x}/g_{COD-s}

4.2.2 Models applied and cases investigated

The system definitions given in the previous section are for the standard case of BM1. Four other cases introduce either oxygen limitation, biomass limitation, or internal or external mass-transfer limitation. The conditions imposed and effects evaluated by each case are summarized in Table 4.4. For all cases, the output variables from the various models include the substrate and oxygen fluxes across the biofilm interface and their concentrations in the bulk liquid, at the biofilm surface, and at the substratum.

The biofilm models included in the BM1 study and the codes by which they are referred to are listed in Table 4.5 and are described in detail in Chapter 3. A first group of biofilm models (codes A, PA, N1a, N1aa, N1s, N2c) provide solutions for a strictly flat biofilm morphology. Models N1s and N2c use multi-dimensional algorithms, but the biofilm is restricted to a flat surface morphology and therefore degenerates to a 1d solution. A second group of biofilm models (NP3a, NP3b, NP3c, N3a, N3b) allows a heterogeneous structure of the biofilm solid matrix, but maintains the same average biofilm thickness as the 1d models. Models N3a and N3b use a true 3d simulation based on Picioreanu *et al.* (1998a). To produce a heterogeneous biofilm morphology, a 3d biofilm solid matrix is grown by these models until the biomass reaches the prescribed value of M_X; then the matrix is frozen and used to calculate the output variables. Figure 4.2 shows a 3d output and illustrates the irregular surface. Two alternative situations with different assumptions for the effective diffusion layer at the biofilm surface are evaluated: N3a assumes that the bulk liquid above the maximum biofilm thickness is completely mixed, while the water in the biofilm channels below the maximum biofilm thickness is stagnant. N3b assumes that the entire water phase (bulk liquid plus the liquid volume in biofilm channels below) is completely mixed.

Table 4.4. Summary of the standard and four special conditions investigated

Case	Name	Condition	Effect evaluated
1	Standard condition	See Section 4.2.1	
2	Oxygen limitation	Bulk liquid oxygen concentration reduced from $S_{O2} = 10\ g_{O2}m^{-3}$ to $S_{O2} = 0.2\ g_{O2}m^{-3}$	Change from substrate limitation to oxygen limitation
3	Biomass limitation	Average biofilm thickness reduced from $L_F = 500\ \mu m$ to $L_F = 20\ \mu m$	Full penetration of the biofilm
4	Reduced diffusivity in the biofilm	Biofilm diffusivity reduced from $D_F = D_L$ to $D_F = 0.2\ D_L$	Internal mass transfer resistance
5	Mass transfer resistance in the bulk liquid	Boundary layer thickness increased from $L_L = 0\ \mu m$ to $L_L = 500\ \mu m$	External mass transfer resistance

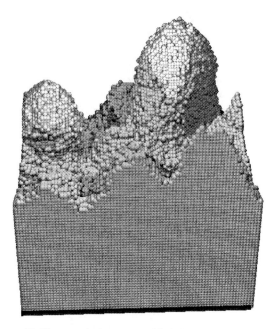

Figure 4.2. Heterogeneous biofilm morphology as used in N3a and N3b.

Models A, PA, and N2c evaluate only the limiting dissolved component (substrate or oxygen) and calculate outputs for the non-limiting compound based on stoichiometry. The other models use dual Monod kinetics for substrate and oxygen to calculate the organic-substrate rate and then use stoichiometry to calculate the outputs for oxygen.

In the pseudo-3d models, NP3a, NP3b, and NP3c, the following assumptions are made to meet the requirement of a L_F = 500 µm: In NP3a and NP3b, the individual biofilm thicknesses $L_{F,i}$ are chosen to be 50, 100, …, 950 µm with constant intervals and w_i = 1/19. In model NP3c, an average biofilm thickness of approximately 500 µm is obtained for the time between detachment events, which is 7.4 d. Immediately after a detachment event, the biofilm thickness is reset to a minimum value of 10 µm.

Table 4.5. Biofilm models applied in the Benchmark 1 study

Code	Type of model	Specification	Reference
A	1d analytical	Combination first/half order kinetics	Pérez *et al.* (2005)
PA	1d pseudo-analytical	Approximation by algebraic equations	Sàez and Rittmann (1992)
N1a	1d numerical	Using AQUASIM	Wanner and Reichert (1996)
N1aa	1d numerical	As N1a, but with diffusivities adjusted	Wanner and Reichert (1996)
N1s	1d numerical	Steady state 1d	Section 3.5
N2c	2d numerical	Cellular automaton	Pizarro *et al.* (2001)
NP3a	Pseudo 3d numerical	AQUASIM with multiple 1d biofilms	Morgenroth *et al.* (2000)
NP3b	Pseudo 3d numerical	As NP3a, but pores completely mixed	Morgenroth *et al.* (2000)
NP3c	Pseudo 3d numerical	As NP3a with dynamic detachment	Morgenroth *et al.* (2000)
N3a	3d numerical	Heterogeneous 3d biofilm solid matrix	Picioreanu *et al.* (1998a)
N3b	3d numerical	As N3a, but pores completely mixed	Picioreanu *et al.* (1998a)

4.2.3 Results for the standard condition (Case 1)

Results for the standard case are shown in Table 4.6. Comparing the substrate flux $j_{F,S}$ and the bulk liquid substrate concentration $S_{B,S}$ shows that all models having a truly flat biofilm (A, PA, N1a, N1aa, N1s, and N2c) yield virtually identical results. The models differ only in some details of model outputs. Models A and PA do not provide information on the substrate concentration profiles inside the biofilm and do not calculate the substrate concentration at the base of the biofilm. Substrate and oxygen concentrations at the base of the biofilm (from models N1a, N1aa, N1s) demonstrate that the standard case has a deep biofilm for the organic substrate, which is the limiting substrate. Oxygen fully penetrates the biofilm with concentrations $S_{0,O2} > 9$ mg$_{O2}$/L, while substrate concentrations at the base of the biofilm are very low. Models A, PA, and N2c require the user to decide *a priori* whether oxygen or organic substrate is limiting. In these models, flux and bulk concentration of the non-limiting component are calculated using stoichiometric relationships between substrate and oxygen utilization. Model N1a represents diffusion in the biofilm differently than in all the other models (Section 2.5.1). Model N1aa is identical to N1a, except that diffusion is modeled as in the other models. The results of N1a and N1aa differ only by a few percent for all cases of BM1.

The effect of mass-transfer resistance in the biofilm pores can be elucidated by comparing models NP3a and N3a, which assume only diffusion in the pores, with models NP3b and N3b, which assume completely mixed conditions in the open pore space. The models with completely mixed pores have much larger fluxes, which are similar to those of a flat biofilm. Having only diffusion in the pores causes a large decrease in flux, as part of the "biofilm volume" has no substrate-consumption activity. The concentration distributions obtained for N3a and N3b are compared in Figure 4.3. For N3a, the lines of equal substrate concentration are more or less parallel to the substratum (Figure 4.3C). This results in low concentrations $C_{LF,S}$ in the valleys, and the average flux into the biofilm, $j_{F,S}$, is 30% lower than the flux of the models assuming a flat biofilm (Table 4.6). The table also shows that very similar results were obtained for model NP3a. The local fluxes calculated by this model are small for sections with thin biofilms and large for sections with biofilms stretching like fingers into the bulk liquid. On the other hand, the flux $j_{F,S}$ of the model N3b, in which substrate concentrations in the pore space are assumed to be identical to the bulk phase concentration, is 6% higher than the flux calculated by models assuming a flat biofilm surface (Table 4.6).

Table 4.6. Values[1] of the output variables calculated for the standard case (Case 1)

Variable	Units	A	PA	N1a	N1aa	N1s	N2c	NP3a	NP3b	NP3c	N3a	N3b
$S_{B,S}$	gCODm^{-3}	4.6	4.3	4.6	4.4	4.4	4.4	**12.1**	4.6	4.8	**12**	**2.9**
$S_{LF,S}$	gCODm^{-3}	4.6	4.3	4.6	4.4	4.4	4.4	**3.2**	4.6	4.8	**0.7**	**2.9**
$S_{0,S}$	gCODm^{-3}			0.004	0.006	0.01	0.01	**0.13**	**0.48**	**0.66**	0.004	0.01
$S_{B,O2}$ [2]	gO2m^{-3}	*10*	*10*	*10*	*10*	*10*	*10*	*10*	*10*	*10*	*10*	*10*
$S_{LF,O2}$	gO2m^{-3}	10	10	10	10	10	10	**8.3**	10	10	**8.1**	10
$S_{0,O2}$	gO2m^{-3}			9.1	9.2	9.2		**7.8**	9.2	9.2	**7.9**	**9.5**
$j_{F,S}$	gCODm^{-2}d^{-1}	5.1	5.1	5.1	5.1	5.1	5.1	**3.6**	5.1	5.0	**3.5**	**5.4**
$j_{F,O2}$	gO2m^{-2}d^{-1}	1.9	1.9	1.9	1.9	1.9	1.9	**1.3**	1.9	1.9	**1.3**	**2.0**
$L_{F,min}$	μm	500	500	500	500	500	500	**50**	**50**	**10**	**340**	**340**
$L_{F,max}$	μm	500	500	500	500	500	500	**950**	**950**	**813**	**840**	**840**
$L_{F,avg}$	μm	500	500	500	500	500	500	500	500	**522**	500	500

[1] Values that are significantly different within a row are in **boldface** and discussed in the text.
[2] The bulk liquid oxygen concentration (*italic*) is a predefined parameter and is not an output variable of the model

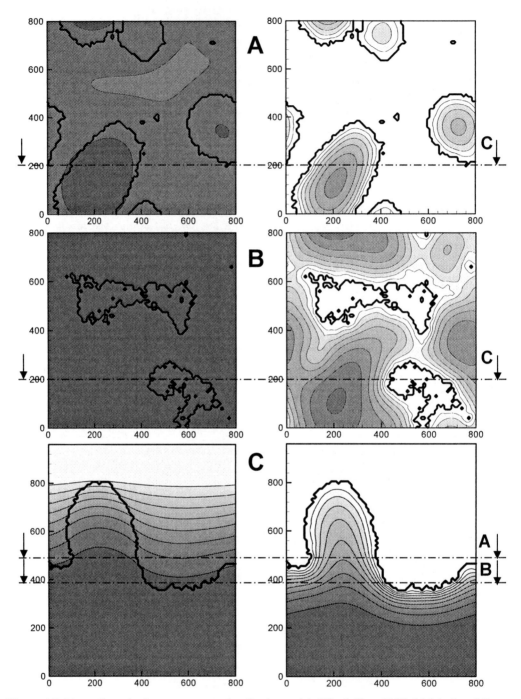

Figure 4.3. Lines of equal substrate concentration for the models N3a (left) and N3b (right). **A** and **B** are horizontal sections in parallel to the substratum at a distance from the substratum of 390 and 490 μm, respectively. **C** is a section perpendicular to the substratum. The white color corresponds to a concentration of 12 g_{COD}/m^3, the darkest grey to 2.3 g_{COD}/m^3. The grey scales in between represent concentration differences of 10%.

This finding is explained by the increased roughness and enlargement of the biofilm surface, which is an essential characteristic of model N3b. Model NP3b has no increase of $j_{F,S}$, as the pseudo 3d approach does not consider mass transport parallel to the substratum and therefore does not capture the effect of surface enlargement. The effect of the enlarged surface area depends on the penetration depth of the limiting substrate compared to the size of biofilm heterogeneities, as will be further discussed in Cases 2 and 3 of BM1.

The issue of spatial versus temporal heterogeneity is investigated by models NP3b and NP3c. Model NP3b evaluates the influence of a truly steady state biofilm thickness distribution, while model NP3c investigates the influence of local variations of biofilm thickness as a result of dynamic local detachment events. The average substrate flux $j_{F,S}$ and the bulk liquid substrate concentration $S_{B,S}$, demonstrate that variation of biofilm thicknesses ranging from 10 to 950 µm do not have a significant influence on the overall performance. The local fluxes for models NP3b and NP3c are shown in Figure 4.4; for $L_F > 200$ µm, the substrate flux becomes independent of the biofilm thickness. The local flux for NP3a (diffusion only in the pores) shows a completely different pattern: thin portions of biofilm have a very low local flux, since they see a low substrate concentration (Figure 4.4).

The effect of dynamic detachment is studied using model NP3c. Figure 4.5 is an example of the development in time of the local biofilm thickness and the local substrate flux. Directly following a detachment event, the substrate flux significantly decreases to 0.5 $g_{COD}m^{-2}d^{-1}$. As the biofilm thickness grows, the substrate flux increases until the biofilm thickness exceeds 200 µm, when substrate flux starts to be limited by mass transport into the thick biofilms.

In conclusion, the standard situation investigated by BM1 shows that the different modeling approaches for a flat biofilm morphology give only minor differences in all model outputs, while different assumptions for mass transfer for rough biofilm surfaces have a major influence on predicted substrate bulk liquid concentrations and fluxes.

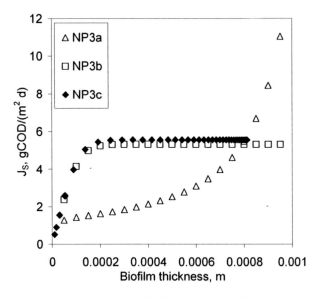

Figure 4.4. Substrate flux into heterogeneous biofilms calculated for steady state biofilm thickness distributions (models NP3a and NP3b) and for dynamic local detachment events (model NP3c).

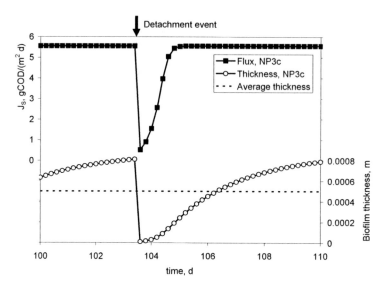

Figure 4.5. Effect of a local detachment event on the development in time of the substrate flux and biofilm thickness. A detachment event is triggered every 7.4 days, decreasing the biofilm thickness down to a base thickness of 10 μm and resulting in an average biofilm thickness of 500 μm.

4.2.4 Results for oxygen limitation (Case 2)

The conditions for Case 2 are identical to Case 1, except for the bulk liquid oxygen concentration, which is reduced from $S_{O2} = 10$ $g_{O2}m^{-3}$ to $S_{O2} = 0.2$ $g_{O2}m^{-3}$. The results are summarized in Table 4.7. Low bulk liquid oxygen concentration shifts the limiting component from the organic substrate to oxygen, as can be seen from the concentrations at the base of the biofilm, $S_{0,S}$ and $S_{0,O2}$. Oxygen concentrations at the base of the biofilm are virtually zero, while the substrate is fully penetrating the biofilm. Models A, PA, and N2c require the user to decide *a priori* which is the limiting component. For a simple scenario like BM1, this decision can be based on ratios of diffusion coefficients and stoichiometric parameters (for details see Chapter 2 in this book and Chapter 5.4 in Henze *et al.* 2000).

Table 4.7. Values[1] of the output variables for oxygen limitation (Case 2)

Variable	Units	A	PA	N1a	N1aa	N1s	N2c	N3a	N3b
$S_{B,S}$	$g_{COD}m^{-3}$	19	19	19	19	19	20	**28**	**13**
$S_{0,S}$	$g_{COD}m^{-3}$			18	18	18	20	**26**	**12**
$S_{B,O2}$ [2]	$g_{O2}m^{-3}$	*0.2*	*0.2*	*0.2*	*0.2*	*0.2*	*0.2*	*0.2*	*0.2*
$S_{0,O2}$	$g_{O2}m^{-3}$			0.0	0.0	0.0		0.0	0.0
$j_{F,S}$	$g_{COD}m^{-2}d^{-1}$	2.3	2.3	2.2	2.2	2.2	2.0	**0.41**	**3.4**
$j_{F,O2}$	$g_{O2}m^{-2}d^{-1}$	0.85	0.84	0.80	0.80	0.82	0.74	**0.15**	**1.3**
$j_{S,rel}$ [3]	%	45	45	43	44	43	39	**12**	**63**
$L_{F,avg}$	μm	360	359	340	360		**500**	350	350

[1] Values that are significantly different are in **boldface** and are discussed in the text.
[2] The bulk liquid oxygen concentration (*italic*) is a predefined parameter and is not an output variable of the model
[3] Ratio of substrate flux $j_{F,S}$ of Case 2 and Case 1.

Modeling results for all models requiring a flat biofilm surface are similar, and substrate fluxes into the biofilm, $j_{F,S}$, are reduced by approximately 56% compared to fluxes calculated for the standard Case 1. The significance of external mass transfer resistance is demonstrated by the models N3a and N3b, which are based on a heterogeneous biofilm morphology. N3a assumes diffusion layer in the pores, and the substrate flux is only 12% of that for Case 1. The lower bulk liquid concentration of oxygen amplifies the effect of the external mass transfer resistance. Model N3b assumes no mass transfer resistance, and the decrease in flux is only 37% from Case 1, because the penetration depth of oxygen is small compared to the biofilm thickness. Thus, Case 2 shows that biofilm morphology and external mass transfer resistance have significant influences on model predictions when the bulk liquid concentration and penetration depth of the limiting substrate are small.

4.2.5 Results for biomass limitation (Case 3)

The conditions for Case 3 are identical to Case 1, except that the average biofilm thickness is reduced from $L_F = 500$ μm to $L_F = 20$ μm. In other words, the reactor contains only 0.02 g_{COD-X} of biomass, as compared to 0.5 g_{COD-X} for Case 1. Comparison of the modeling results reveals that all models except N2c give similar model outputs (Table 4.8), independent of biofilm morphology. Thus, mass transport do not influence overall substrate degradation when the total amount of biomass is limiting. The concentrations at the base of the biofilm document that the biofilm is fully penetrated for organic substrate and oxygen. In general, the advantages of the more complex models are minimized if the biofilm is biomass limited and fully penetrated.

Table 4.8. Values[1] of the output variables for biomass limitation (Case 3)

Variable	Units	A	PA	N1a	N1aa	N1s	N2c	N3a
$S_{B,S}$	$g_{COD}m^{-3}$	21	22	22	22	22	17.2	22
$S_{0,S}$	$g_{COD}m^{-3}$			22	22	22	16.9	22
$S_{B,O2}$ [2]	$g_{O2}m^{-3}$	*10*	*10*	*10*	*10*	*10*	*10*	*10*
$S_{0,O2}$	$g_{O2}m^{-3}$			10	10	10		9.9
$j_{F,S}$	$g_{COD}m^{-2}d^{-1}$	1.9	1.6	1.6	1.6	1.6	**2.6**	1.6
$j_{F,O2}$	$g_{O2}m^{-2}d^{-1}$	0.70	0.58	0.58	0.58	0.58	**0.96**	0.59
$j_{S,rel}$ [3]	%	37	31	31	31	31	**51**	46

[1] Values which are significantly different are in **boldface** and are discussed in the text.
[2] The bulk liquid oxygen concentration (*italic*) is a predefined parameter and is not an output variable of the model
[3] Ratio of substrate flux $j_{F,S}$ of Case 3 and Case 1

The difference in the results of N2c can be attributed to a relatively rough spatial discretization of the biofilm. The size of the numerical grid was maintained at 4 μm, meaning that the biofilm of 20-μm thickness was represented by just 5 grid points, a number insufficient to provide an adequate representation of the spatial gradients within the biofilm. Therefore, model N2c produces an inaccurate solution of the problem in Case 3. Errors are small for N2c when the biofilm is thicker in the other cases.

4.2.6 Results for reduced diffusivity in the biofilm (Case 4)

A 80% reduction of the diffusion coefficients inside the biofilm for the organic substrate and dissolved oxygen results in approximately 20% decreases in fluxes into the biofilm for models assuming the biofilm surface to be flat (Table 4.9). The diffusion coefficient has a modest impact on model predictions. This impact is relatively low because, for a deep biofilm, the substrate flux is proportional to $\sqrt{D_{F,S}}$ and $\sqrt{S_{B,S}}$ (Harremoës 1976). Thus, for a given bulk liquid substrate concentration $S_{B,S}$, a decrease of the diffusion coefficient $D_{F,S}$ by a factor of 5 results in a decrease of the substrate flux into the biofilm by a factor of 2.2. As can readily be seen from an overall mass balance, a decrease of the flux means an increase of the bulk liquid substrate concentration, which in turn increases the substrate flux. Therefore, only a minor net decrease of the flux is caused by the substantial decrease of the diffusion coefficient.

Models N3a and N3b, which assume heterogeneous biofilm morphology, illustrate that the effect of a reduced diffusivity in the biofilm is strongly related to the assumptions made with regard to external mass transfer. For model N3a, in which the water in the biofilm pores below the maximum thickness is stagnant, transport in these pores also is affected by the decrease of the diffusion coefficient. This results in a reduction of the substrate flux by 49% as compared to Case 1 of BM1 (Table 4.9). The reduction is higher than that of the models assuming a flat biofilm surface, because the biofilm thickness of those models is smaller than the maximum biofilm thickness in N3a. Since N3b assumes that the biofilm pore volume is completely mixed, the enlarged surface area of this model becomes more and more of an advantage as mass transport inside the biofilm becomes limiting. Thus, the flux in model N3b decreases only by 11%, compared to Case 1, i.e., the decrease is significantly smaller than that of the models assuming a flat biofilm surface. In conclusion, transport of dissolved components in the biofilm is more affected by the assumptions made with regard to transport in the pore water and by the morphology of the biofilm than by the value of the diffusion coefficient.

Table 4.9. Values[1] of the output variables for reduced diffusivity in the biofilm (Case 4)

Variable	Units	A	PA	N1a	N1aa	N2c	N3a	N3b
$S_{B,S}$	$g_{COD}m^{-3}$	9.2	9.3	9.8	9.4	9.7	**22**	**6.1**
$S_{0,S}$	$g_{COD}m^{-3}$			0	0	0	**0.42**	0
$S_{B,O2}$ [2]	$g_{O2}m^{-3}$	*10*	*10*	*10*	*10*	*10*	*10*	*10*
$S_{0,O2}$	$g_{O2}m^{-3}$			8.2	8.3		**0**	**4.9**
$j_{F,S}$	$g_{COD}m^{-2}d^{-1}$	4.2	4.1	4.0	4.1	4.1	**1.8**	**4.8**
$j_{F,O2}$	$g_{O2}m^{-2}d^{-1}$	1.5	1.5	1.5	1.5	1.5	**0.56**	**1.8**
$j_{S,rel}$ [3]	%	82	80	78	80	80	**51**	**89**

[1] Values that are significantly different are in **boldface** and are discussed in the text.
[2] The bulk liquid oxygen concentration (*italic*) is a predefined parameter and is not an output variable of the model
[3] Ratio of substrate flux $j_{F,S}$ of Case 4 and Case 1

4.2.7 Results for external mass transfer resistance (Case 5)

Case 5 introduces external mass transfer resistance in the form of a liquid boundary layer of thickness $L_L = 500$ μm. For the models requiring a flat biofilm, the boundary layer is a stagnant zone just above the biofilm surface. For model N3a, assuming a heterogeneous biofilm matrix, the layer reaches from the maximum biofilm thickness into the bulk liquid. For model N3b, which also assumed the heterogeneous biofilm morphology, the layer follows the contour of the biofilm surface.

Model N3b and the models having a flat biofilm give very similar fluxes (Table 4.10), which were about 45% less than for Case 1. For model N3a, the reduction of the flux relative to that of the standard case is only 34%. As N3a assumes an external mass transfer resistance in the pore space of the biofilm for all the five cases, the substrate flux of Case 1 is already smaller than that of the other models. Thus, the relative flux $j_{S,rel}$ for N3a for Case 5 is larger than that of the other models, even if $j_{F,S}$ of N3a is smaller. Hence, external mass transfer resistance can be a very significant, depending on the thickness of the stagnant zone. When external mass transport is significant, biofilm morphology is not of primary importance.

Table 4.10. Values[1] of the output variables calculated for mass transfer in the bulk liquid (Case 5)

Variable	Units	A	PA	N1a	N1aa	N2c	N3a	N3b
$S_{B,S}$	$g_{COD}m^{-3}$	16	16	16	15.4	15.7	**19**	16
$S_{LF,S}$	$g_{COD}m^{-3}$	2.5	2.1	2.2	2.2	2.3	**7.5**	1.4
$S_{0,S}$	$g_{COD}m^{-3}$			0.005	0.005	0.018	**0.005**	0.005
$S_{B,O2}$[2]	$g_{O2}m^{-3}$	*10*	*10*	*10*	*10*	*10*	*10*	*10*
$S_{LF,O2}$	$g_{O2}m^{-3}$		7.4	7.4	7.6		7.9	7.4
$S_{0,O2}$	$g_{O2}m^{-3}$			7	7		5.9	7.1
$j_{F,S}$	$g_{COD}m^{-2}d^{-1}$	2.8	2.8	2.8	2.8	2.9	**2.3**	2.9
$j_{F,O2}$	$g_{O2}m^{-2}d^{-1}$	1.0	1.0	1.0	1.0	1.1	**0.83**	1.1
$j_{S,rel}$[3]	%	55	55	55	55	57	**66**	54
$L_{F,avg}$	μm	430		440	461	500	495	495

[1] Values that are significantly different are in **boldface** and are discussed in the text.
[2] The bulk liquid oxygen concentration (*italic*) is a predefined parameter and is not an output variable of the model
[3] Ratio of substrate flux $j_{F,S}$ of Case 5 and Case 1

4.2.8 Lessons learned from BM1

For the simple conditions of BM1, modeling results are not significantly different for all modeling approaches having a flat biofilm morphology. On the other hand, modeling results are strongly influenced by the assumption for mass transfer in the pores within the biofilm when a heterogeneous biofilm morphology is allowed. If mass transfer within the biofilm pore volume, i.e., in the liquid phase below the maximum biofilm thickness, is by diffusion only, overall fluxes decrease significantly. With the assumption of concentrations within the biofilm pore volume being equal to the bulk phase concentration, the enlarged surface area results in a slight increase of fluxes of dissolved components into the biofilm. This enlarged surface area is an advantage as long as the penetration depth is small compared to the roughness of the heterogeneous biofilm. The influence of biofilm morphology and external mass transfer resistance are discussed further in BM2.

While modeling results are similar for most modeling approaches, the effort in implementing and using the different models is not. Models A and PA, based on analytical or pseudo-analytical solutions, respectively, can be readily solved using a spreadsheet. However, analytical or pseudo-analytical solutions require a number of simplifications, and the modeler has to make *a priori* decisions, e.g., on the dissolved component that is rate limiting. Numerical 1d models can be solved using readily available software (models N1a and N1aa), e.g., the simulation package AQUASIM (Reichert 1998a, 1998b). These simulations can be performed on a PC, and the modeling results are available within minutes. To approximately evaluate the influence of a heterogeneous morphology, 1d simulations can be combined in pseudo-3d models (NP3a, NP3b, NP3c). Multi-dimensional models are able to simulate heterogeneous biofilm morphology (N3a, N3b), but they require custom-made software and, in some situations, more extensive computing power.

Thus, for simple biofilm systems and more-or-less smooth biofilm surfaces, analytical, pseudo-analytical, or simple numerical 1d models often provide good compromise between the required accuracy of modeling results and the effort involved in producing these results. Adopting the more complex and intensive 2d and 3d models is justified only when the heterogeneity that they allow is critical to the modeling objective.

4.3 BENCHMARK 2: INFLUENCE OF HYDRODYNAMICS

The second benchmark study involves spatially heterogeneous architectures that can induce complex flow patterns and affect mass transfer (Eberl *et al.* 2000a; Picioreanu *et al.* 2000a, 2000b). Classical 1d biofilm models are not able to capture this kind of complexity, which historically has been one of the reasons for the development of multi-dimensional models.

Specifically, the assumption of a completely mixed bulk fluid is given up in this benchmark problem, and mass transport due to diffusion and advection in the fluid compartment are explicitly considered. The latter implies that the hydrodynamic flow field should be taken into account as well. A direct microscopic mathematical description (cf. section 3.1) leads to a nonlinear system of 3d partial differential equations in a complicated domain, and this is numerically expensive and difficult to solve. Therefore, we investigate to what extent such a detailed local description of physical and spatial effects is necessary for macro-scale applications, where the purpose of the modeling is often only to calculate the total mass fluxes into the biofilm, *i.e.,* the global mass conversion rates. To this end, the description of the physical complexity of the system, expressed in geometrical and hydrodynamic complexity, can be simplified in various ways and to varying degrees in order to obtain faster simulation methods. The results of these simplified models are compared with the results of a fully three-dimensional simulation study.

4.3.1 Definition of the system modeled

BM2 picks up on a mass-transfer issue that has been studied previously in several investigations (e.g., Rittmann and McCarty 1980, Rittmann *et al.* 1999, Picioreanu *et al.* 2000a, Eberl *et al.* 2000a). While each of these earlier works focused on one class of models, we conduct a comparative analysis of several approaches. Based on the usual time-scale argument (cf. Section 2.6.3) that growth of biofilms is much slower than the transport processes for the dissolved substrate, the system is considered to be in a frozen steady state. This means that we assume the mesoscopic biofilm structure to be given and not changing over time. Hence, we are looking at a small time window, one long enough for the transport processes to relax to equilibrium, but short compared to the characteristic time scale of biofilm formation.

The physical processes considered are mass transport of one substrate due to diffusion in the solid region (that is, the region with a positive biomass density) and due to advection and diffusion in the liquid region (that is, the region with no biomass). Advective transport is created by a hydrodynamic flow field. For BM2, we assume that advective transport is driven by a Couette-like mechanism, i.e., the top boundary of the domain is moving at a specified velocity. Imposing periodic boundary conditions on the flow equations mimics a self-repeating biofilm architecture. This model assumption is discussed, e.g., in Picioreanu et al (2000a), Eberl et al (2000a), and Sudarsan et al (2005).

The biomass density $X_F = X_F(x,y,z)$ is assumed to be constant throughout the biofilm, i.e., no biomass gradients. This agrees with previous *model assumptions* in mass-transfer studies in irregular biofilm structures (Rittmann *et al.* 1999, Picioreanu *et al.* 2000a, Eberl *et al.* 2000a) and in some dynamic biofilm-formation models, such as Hermanowicz (2001), Mehl (2001), Dockery and Klapper (2001), and with *model results* for the special case of one species (Noguera *et al* 1999) and for the special cases of neglectible biomass decay due to starvation (Eberl *et al* 2001; Efendiev *et al* 2002). In the biofilm, substrate is consumed according to the usual Monod kinetics.

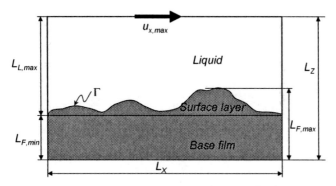

Figure 4.6. Schematic of the biofilm system in BM2: an irregular biofilm is divided into a base film that determines its thickness and a surface layer that determines its roughness. The flow is driven by a plate on the top that moves at constant velocity $u_{X,max}$, i.e., by a Couette mechanism.

Due to the high computational demand of 3d fluid dynamics in irregular domains, the problem is restricted to a small section of a biofilm (1.6 mm long). The size of the rectangular channel segment is $L_X \times L_Y \times L_Z$. The direction of primary flow is denoted x, and the vertical direction is z (see Figure 4.6), leaving y for the secondary flow direction. Internally, this segment is divided into two regions, one containing the actual biofilm, the other one containing water only.

The goal of BM2 is to calculate: (1) the flux of dissolved substances into the solid region; and (2) the average substrate concentrations at the solid/liquid interface and at the substratum. A crucial aspect of the formulation of BM2 is the specification of appropriate boundary conditions (cf. Section 3.1.1.2). These are required to connect the small computational system with the external world that surrounds the modeling domain. The specific boundary domain determines the unique model solution, *i.e.*, different boundary conditions give different solutions of the underlying model equations. The inflow condition imposed on the substrate concentration field S is

$$u(0, y, z) \cdot S(0, y, z) - D \cdot \frac{\partial S(0, y, z)}{\partial x} = u(0, y, z) \cdot S_B \qquad (4.1)$$

A no-flux condition is imposed on all other boundaries of the domain.

The results of the various model simulations are measured and compared using two lumped performance criteria that can be calculated by all methods.

Total flux/biomass production. In steady state, the total production of biomass $r_{A,pro,X}$ ($M_X L^{-2} T^{-1}$) is equivalent to the total flux $j_{F,S}$ of dissolved substrate S from the liquid into the solid biofilm ($M_S L^{-2} T^{-1}$)

$$Y \cdot j_{F,S} A_F = r_{A,pro,X} A_S \qquad (4.2)$$

that is, all substrate is converted into new biomass. The quantities are defined by

$$j_{F,S} = D_L \int_\Gamma \partial_n S \, dA \qquad (4.3)$$

$$r_{A,pro,X} = \frac{1}{A_S} \int_{V_F} \mu_{max} Y \frac{S}{K+S} X_F \, dV \qquad (4.4)$$

where Γ denotes the liquid-biofilm interface with an area A_F, A_S is the biofilm substratum area, and $\partial_n S$ is the gradient of substrate concentration normal to the biofilm surface Γ. It is, therefore, sufficient to restrict oneself to one of these two parameters. We will use $j_{F,S}$ in our study.

Average surface concentration. The flux $j_{F,S}$ describes how much substrate is transported into the biofilm, but it does not contain quantitative information about the substrate concentration. In order to describe this, the average surface concentration ($S_{LF,avg}$) is evaluated as a second performance index.

$$S_{LF,avg} = \frac{\int_\Gamma S\, dA}{\int_\Gamma dA} = \frac{\int_\Gamma S\, dA}{A_e\, L_X\, L_Y} \tag{4.5}$$

The criteria $j_{F,S}$ and $S_{LF,avg}$ describe the performance of the biofilm system from a macro-scale point of view. Thus, comparing these values for different modeling approaches addresses the question of whether or not a finer resolution of the system, i.e., carried out in 2d and 3d approaches, is necessary if only the macro-scale behavior of the reactor is of interest, as it is the case in many engineering applications. On the other hand, only the 2d and 3d numerical models can predict the micro-scale detail, which may be the modeling goal in certain research applications.

4.3.2 Cases investigated

In total, 15 different scenarios are compared in BM2. They vary in biofilm thickness (thin versus thick biofilms), biofilm surface architecture (flat, wavy, and "bumpy" cluster-ad-channel morphologies), flow velocity (fast versus slow), and (in one case only) bulk substrate concentration. The surface morphology of the biofilms is specified as input data. Three different values for the base-film thickness are investigated in the case of the wavy biofilm and two different values for each of the other architectures. Similarly, two different flow velocities are taken into account for each of the three biofilm surface architectures. The bulk concentration of the dissolved substrate is kept at the same value for all but one test case, where it is reduced to 10% of its original value. The length of the biofilm segment is 1.6 mm for every scenario, while width and height vary. All other constants, including Monod reaction parameters, hydrodynamic constants, and diffusion coefficients, are not varied in this simulation experiment.

All unchanged model parameters are listed in Table 4.11, and cases simulated in BM2 are summarized in Table 4.12.

Table 4.11. List of parameters that remain unchanged for all simulation experiments

Physical quantity	Variable	Value
Length of system	L_X	1.6 mm
Kinematic viscosity	ν	0.00864 m^2/d
Fluid density	ρ	1 g/d
Biomass density	X_F	10000 g/m^3
Diffusion coefficient	D	10^{-4} m^2/d
Monod constant	K	4 g/m^3
Maximum specific growth rate	μ	5.88 1/d
Yield coefficient	Y	0.63 g$_{COD-X}$/g$_{COD-S}$

Table 4.12. Overview of cases in BM2. The first five columns display some characteristics of the biofilm structures, while the last two columns describe environmental conditions. The cases are grouped with respect to the biofilm type: **F** flat homogeneous films , **W** wavy biofilm morphology, or **B** bumpy cluster-and-cell architecture.

Case	Height of system L_Z [mm]	Base film thickness, $L_{F,min}$ [mm]	Avg. biofilm height L_F [mm]	Max biofilm height, $L_{F,max}$ [mm]	Volume of Biofilm, V_F [10^{-12} m^3]	Bulk concentration S_B [g/m^3]	Flow velocity $u_{X,max}$ [m/d]
F1	0.3765	0.0628	0.0628	0.0628	3.15	5.0	864.0
F2	0.6275	0.3137	0.3137	0.3137	15.80	5.0	864.0
F3	0.3765	0.0628	0.0628	0.0628	3.15	5.0	86.4
F4	0.6275	0.3137	0.3137	0.3137	15.80	5.0	86.4
W1	0.3953	0.1067	0.1358	0.1694	20.50	5.0	140
W2	0.3953	0.1067	0.1358	0.1694	20.50	5.0	14
W3	0.5835	0.2949	0.3241	0.3576	48.80	5.0	140
W4	0.5835	0.2949	0.3241	0.3576	48.80	5.0	14
W5	0.3012	0.0126	0.0417	0.0753	6.29	5.0	140
W6	0.3012	0.0126	0.0417	0.0753	6.29	5.0	14
W7	0.3953	0.1067	0.1358	0.1694	20.50	0.5	140
B1	0.3514	0.0126	0.0381	0.0753	9.48	5.0	8.64
B2	0.5647	0.0226	0.2442	0.2886	75.90	5.0	8.64
B3	0.3514	0.0125	0.0381	0.0753	9.48	5.0	43.2
B4	0.5647	0.0226	0.0244	0.2886	75.90	5.0	43.2

4.3.3 Models applied

Six models are used in BM2, ranging from a fully 3d simulation code to simplified one-dimensional approaches. Table 4.13 provides a summary overview, more detailed descriptions of the individual models are given in the subsequent sections.

Model N1a is included only in some parts of the study where it plays the role of a test group. Since by construction it does not depend on hydrodynamics, it should not be sensitive to flow conditions. At his point, the 1d parameter fitting approaches (A, PA, N1, N1b-d) are not closed, but require the results of (N3c); thus they do not provide a predictive solution of BM2.

4.3.3.1 Three dimensional model (N3c)

The most comprehensive, albeit most computing-intensive approach to solve BM2 is to solve the complete 3d steady-state model. It consists of the diffusion-reaction equation (3.12) in the biofilm and of the advection-diffusion reaction (3.9) with $r=0$ in the fluid compartment. The latter requires also the numerical solution of the Navier-Stokes equations (3.8) in the irregular domain that is defined by the biofilm morphology. Across the biofilm/water interface, the concentration field and the flux must be continuous. With these interface conditions, the advection-diffusion/diffusion-reaction problem is well-posed (in the weak sense). From the numerical solution $S(x,y,z)$, the flux $j_{F,S}$ and then average concentration at the biofilm/water interface, $S_{LF,avg}$, can be computed. Since this model does not introduce any reduction of complexity it is the most detailed model in the survey, and its results are used as the reference for the other approaches.

Table 4.13. Models used in the BM2

Model code	Brief description	Reference
N3c (3d, full hydrodynamics)	A fully 3d numerical simulation of hydrodynamics and mass transfer in the specified biofilm architectures (the original problem; serves as reference for comparison)	Eberl *et al* (2000a)
N2b (2d, full hydrodynamics)	A reduction of the 3d biofilm architecture to a 2d one with comparable characteristic geometrical property, along with full 2d hydrodynamics and mass transfer.	Picioreanu *et al* (2000a)
N2d, N2e (2d, simplified hydrodynamics)	As in N2b, but further simplifying the hydrodynamics, not taking the irregularity of the structure into account for flow field calculations; two different 2d reductions of the biofilm are used, denoted by N2d and N2e	This section and section 3.6.4.6
N1sc (1d, *a priori* Sh correlation)	A 1d model that does not resolve the flow field. The liquid phase is divided into a completely mixed bulk region and a concentration boundary layer. Describing the effect of hydrodynamics on mass transfer requires an additional model parameter, the concentration boundary layer thickness L_l. This parameter is not given, but must be computed. Model approach N1sc derives it from an *a priori* correlation for mass transfer based on Comiti *et al.* 2000.	This section
N1s (1d, global mass balance)	A 1d description of the biofilm; hydrodynamics is incorporated by a global mass balance for the system (cf. 1d steady-state model). This approach can be understood as being similar to N1sc, but it uses a different method to determine the concentration boundary layer.	Section 3.5
N1a (1d without hydrodyanmics)	A 1d approach; the biofilm is divided into 2 characteristic regions according to Morgenroth and Wilderer (2000). Instead of explicit hydrodynamics an arbitrarily fixed boundary layer is used, i.e., variations in the hydrodynamics are not taken into account.	Morgenroth and Wilderer (2000)
A, PA, N1, N1b-d (1d parameter fitting)	Reduction to a lumped 1d system with completely mixed bulk and concentration boundary layer; as in (N1sc), variations in the hydrodynamics are expressed in terms of the concentration boundary layer thickness L_l. It is determined by parameter fitting in order to establish the correlation between L_l and the flow conditions. The results of (N3c) are used as reference. The parameter-fitting approach is not unique. Several variants are compared, denoted by (N1, N1b-d, PA, A).	This section and sections 3.2-3.4

The numerical method used for the flow-field simulations is described in Eberl *et al.* (2000a). In order to calculate the substrate concentration, a standard cell-centered finite-volume upwind scheme is used on a regular grid. The size of a computational cell is the size of a voxel in the description of the biofilm geometry. This numerical method is a standard scheme and the easiest one to implement. It is used in many off-the shelf software packages, included in many numerical libraries, and described and discussed in many introductory textbooks on numerical treatment of partial differential equations (e.g., Morton, 1996). Its simplicity does not come without drawbacks, though. In order to allow a statement on the accuracy of the simulation, we compare the flux of substrate into the solid biofilm with the flux from the liquid phase. In the cases of rough biofilm geometry with a thin base layer, this mass balance criterion cannot be satisfied with the desired accuracy (cases 1 and 3). Hence, the chosen discretization (governed by the provided data) was too coarse in these cases. In all other cases, the mass-balance criterion yields satisfactory results. This observation for rough biofilms illustrates why is it essential to check for mass-balance closure when using any numerical method for solving partial differential equations.

A typical 3d simulation shows that the substrate concentration decreases in the main flow direction and towards the substratum. That is, the highest concentration values are at the inflow, close to the top boundary of the domain, while the lowest concentration values are at the outflow close to the substratum. This is the obvious correct trend due to substrate consumption in the biofilm. Independent of the type of geometry, the substrate concentration in the channel segment is higher when flow field moves faster. This is depicted in Figure 4.7, which shows simulation results for irregular biofilm architectures. Plotted are iso-surfaces of $0.99 \cdot S_B$ for two different velocities per biofilm geometry. In both cases, the iso-surface belonging to the slower flow velocity lies above the one for the faster velocity, indicating smaller concentration values. The iso-surfaces in Figure 4.7 do not express spatial variations in y direction. This shows that the spatial variation of the biofilm surface in this secondary direction do not propagate and amplify in the concentration field in the liquid region; to the contrary, the variations weaken. This is an expected consequence of the smoothing property of Fickian diffusion. From the macro-scale point of view, the concentration field in the liquid region appears almost two-dimensional, where the variations in primary flow direction x are due to advection and depletion.

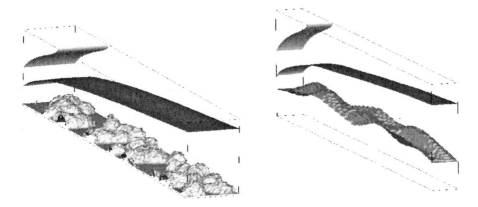

Figure 4.7. The bumpy (left) and the wavy (right) biofilm surface with simulation results: plotted are the 99% iso-surfaces, i.e. the surfaces where $S=0.99 \cdot S_B$, for two simulations each (Cases B1, B3, W1, W3). The iso-surface is closer to the substratum if the flow velocity in the bulk is higher, implying higher concentration values and a higher mass transfer. Furthermore, these iso-surfaces do not show variations in the secondary flow direction y.

4.3.3.2 Two-dimensional models

4.3.3.2.1 Reduction to a 2d description (N2b)

The 99% iso-surfaces shown in Figure 4.7, as well as discussion in Eberl *et al* (2000a), suggest that a reduction of the 3d problem to a 2d description will work well if only global (macro-scale) results are of interest. Going to 2d means giving up the secondary flow direction y. If the discretization accuracy is maintained, the number of unknown grid variables in the numerical computation is reduced by a factor that equals the number of grid points in that direction. Moreover, computation of a grid variable in a 2d setup requires fewer mathematical operations than a 3d setup. For example, in a Lattice-Boltzmann method, 9 discrete directions of particle movement must be considered in 2d, while at least 15 are needed for 3d. Standard upwind/cell centered Finite Difference/Volume/Element methods lead to matrices with 5 diagonals in 2d and 7 diagonals in 3d. High Order Compact Stencil methods lead to matrices with 9 diagonals in 2d and to matrices with 19 diagonals in 3d. Thus, a matrix-vector product is much more costly in 3d than in 2d, even if the vector length is the same. Since matrix-vector products are among the most computing expensive parts of a simulation code for partial differential equations, 2d simulations perform much faster than 3d ones, even if some of the computing time that is gained by this reduction is invested in a finer or more accurate discretization of the underlying partial differential equations. In fact, modern personal computers are suitable for most 2d studies, while fully 3d studies of hydrodynamics are very demanding for a PC in regard to memory and CPU performance.

It is not possible to describe a fully 3d biofilm architecture in a 2d way without loss of information. Moreover, converting a 3d biofilm morphology to a 2d description does not follow a simple, fixed recipe. The first and simplistic approach is to average the biofilm height over y. This strategy maintains the volume of the biofilm and, hence, the total amount of biomass, which is an important property for the purpose of comparison. However, one can easily see that this leads to a biofilm description with a maximum height lower than the original three-dimensional architecture, while the minimum height is higher. Consequently, the resulting 2d biofilm is smoothed. On the other hand, Picioreanu *et al.* (2000a) demonstrated that the area enlargement number (Ae) of a biofilm architecture is a good dimensionless descriptor of biofilm irregularity from the global mass transfer point of view. Ae is the ratio of the biofilm/liquid interface and the substratum area covered by the biofilm. Moreover, Eberl *et al.* (2000a) showed that 2d and 3d biofilm descriptions with the same biomass volume and the same area enlargement number lead to quantitatively similar values for the total flux $j_{F,S}$ into the biofilm. This observation suggests using a 2d reduction of the biofilm architecture that, in addition to the total volume, also maintains the area enlargement number Ae. The 2d biofilm geometry appears visually rougher than a 3d one with the same value Ae, a fact that follows directly from the definition of the area enlargement factor.

4.3.3.2.2 Simplification of the hydrodynamic description

The numerical simulation of the Navier-Stokes equations (3.8) is more time consuming than the solution of the convection-diffusion-reaction model that describes mass transfer and biochemical reactions. Hence, a simplified description of the hydrodynamics in the 2d biofilm system should greatly reduce computing demands.

Hydrodynamics affects meso-scale biofilm processes mainly in two ways: transport of substrates and detachment. Since the latter effect is not considered in BM2, it does not pose any restrictions for simplifications in our context. Moreover, one of the conclusions of Eberl *et al.* (2000a) was that the flow field in the channels and pores does not remarkably contribute to an increased net mass conversion in the biofilm from the global point of view. On the other hand, the inflow boundary conditions for the substrate (Equations (3.9) and (3.12)) show how essential hydrodynamics is for the supply of substrate.

An irregular biofilm surface can be viewed as a perturbation of a flat biofilm that, under laminar conditions, induces a linear Couette flow profile

$$u(x, y, z) = u_{x,\max} \frac{z - L_{F,min}}{L_Z - L_{F,min}} \quad \text{for} \quad L_{F,min} \le z \le L_Z \qquad (4.6)$$

Since the local flow velocities in the pores and channels of the biofilm and close to the biofilm/liquid interface are small compared to the flow velocities in the bulk region, it seems intriguing to use, as a first approximation, the Couette flow profile (4.6) as an approximation to the Navier-Stokes equations, even in the irregular domain (see Figure 4.8). It should be noted, however, that superimposing the Couette-flow profile on the irregular surface biofilm layer introduces advective transport inside the biofilm (which does not satisfy (3.12)) and, hence, induces increased mass transfer.

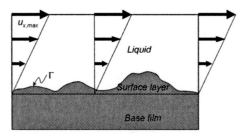

Figure 4.8. Simplified description of hydrodynamics in N2d and N2e: the linear Couette flow profile is used as an approximation to the Navier-Stokes equations.

4.3.3.2.3 Two-dimensional models N2b, N2d and N2e

Three 2d models are used to solve BM2. Model N2b solves the 2d model equations in an *Ae*-preserving 2d biofilm geometry with full hydrodynamics. Lattice-Boltzmann schemes are used for the discretization of fluid flow and mass transfer equations. Models N2d and N2e uses the simplified hydrodynamic model. It is applied to the same *Ae*-preserving geometry as N2b and to the 2d biofilm description that is obtained by averaging the biofilm height along the secondary direction *y*.

In the case of the flat and homogenous biofilm architecture, all 2d and 3d descriptions coincide, and differences in the simulation results are only due to the choice of the numerical schemes and the method parameters controlling their performance.

4.3.3.3 One-dimensional models

4.3.3.3.1 One-dimensional biofilm descriptions and concentration boundary layers

Two possibilities exist to further reduce the model to a 1d description: (i) lumping over the primary flow direction, (ii) or lumping perpendicular to the substratum, *i.e.* averaging (in some sense) over the preferred direction of diffusive transport into the biofilm. The first approach is consistent with classical biofilm models and is the only one considered here. Thus, all 1d models rely on a reduction to a homogeneous slab description of the biofilm, as in Figure 4.9. The key concept is to divide the computational system into three layers: the solid biofilm, in which all the biomass is concentrated; a concentration boundary layer (of *a priori* unknown thickness, located in the liquid compartment); and the bulk liquid. In the latter, the substrate concentration is assumed to be constant. The boundary layer and the solid biofilm are governed by 1d versions of Eqs. (3.9) and (3.12), respectively. Since advective transport only occurs in the main flow direction, *i.e.* parallel to the layers, it does not contribute to mass transfer into the solid biofilm and can be neglected in the boundary layer. Hence, the 1d model for solid biofilm and boundary layer reads

$$\begin{cases} \dfrac{\partial^2 S}{\partial z^2} = 0, \ L_F \le z \le L_F + L_L \\[2mm] \dfrac{\partial^2 S}{\partial z^2} = -\dfrac{\mu_{max} X_F}{YD} \dfrac{S}{K+S}, \ z \le L_F \\[2mm] S(L_F + L_L) = S_B, \ \left.\dfrac{\partial S}{\partial z}\right|_{z=0} = 0 \end{cases} \qquad (4.7)$$

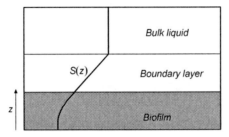

Figure 4.9. In a one-dimensional representation, the domain is layered in three compartments, the solid biofilm in which the substrate concentration is governed by a diffusion-reaction equation, the boundary layer in which the substrate concentration profile is linear, and the bulk in which the substrate concentration is constant. The thickness of the concentration boundary layer enters the one-dimensional models in this study as a model parameter. It is not known *a priori*.

In equations (4.7), L_F denotes the height of the solid biofilm (biofilm thickness) and L_L denotes the thickness of the boundary layer (L). In the context of BM2, it can be assumed that L_F is a known input parameter to be calculated from the biofilm descriptions, while L_L is not known *a priori* from the problem specification. In order to preserve the total amount of biomass, it is clear that $L_F = L_{F,avg}$, the average height of the biofilm.

In the boundary layer, the differential equation is solved analytically: A linear profile is obtained connecting the values S_B (specified as an input parameter) and S_{LF} (not known *a priori*), i.e.,

$$S(z) = (S_B - S_{LF})\dfrac{z - L_F}{L_L} + S_{LF}, \quad L_F \le z \le L_F + L_L \qquad (4.8)$$

In the solid biofilm region, a two-point boundary value problem must be solved. The solution for the boundary layer is included in the equation for the solid biofilm region as:

$$\begin{cases} \dfrac{\partial^2 S}{\partial z^2} = -\dfrac{\mu_{max} X_F}{YD} \dfrac{S}{K+S}, \quad \text{for} \ z \le L_F, \\[2mm] S_{LF} + L_L \left.\dfrac{\partial S}{\partial z}\right|_{z=LF} = S_B, \ \left.\dfrac{\partial S}{\partial z}\right|_{z=0} = 0 \end{cases} \qquad (4.9)$$

Thus, the *a priori* unknown thickness of the boundary layer L_L enters this model as an additional model parameter through the boundary conditions. So far, no information about the hydrodynamic condition is considered explicitly in this one-dimensional model. Implicitly it is included in the concept of the boundary layer, which becomes thinner as the bulk water velocity or turbulence increase (see also Rittmann *et al.* (1999) for an application of this idea in a multi-dimensional set-up, as well as Figure 4.7, which shows how faster bulk flow implies that the 99% iso-surface is closer to the biofilm). Thus, the question posed by the one-dimensional description is: Is there a value L_L for the boundary layer thickness that correctly represents the hydrodynamics in a spatially heterogeneous biofilm architecture (in the sense of $j_{F,S}$ and S_{LF}), and, if so, how is a good estimate obtained?

In order to explore the sensitivity of the wanted quantities flux J and surface concentration S_{LF} on the new model parameter L_l, simulations are carried out for various values of L_l and varying average biofilm thickness L_F, using the diffusion-reaction parameters of BM2's specification. Figure 4.10 shows that the sensitivities of flux and surface concentration decrease for sufficiently large values of L_F and L_l, which corresponds to thick films and slow hydrodynamics in the bulk. Thin biofilms or fast flow fields (i.e., thin boundary layers) give steep variations in flux and substrate concentration. Hence, it is paramount to have an accurate estimate for the boundary layer thickness in order to represent hydrodynamic effects correctly when the boundary layer is thin.

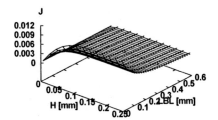 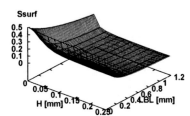

Figure 4.10. Substrate flux [g/d] into the biofilm (left) and substrate concentration [g/m³] at the interface (right) as a function of biofilm height L_F and boundary layer thickness L_l. Both quantities respond very sensitively to variations thin biofilms and thin concentration boundary layers. On the graphs H=L_F, LBL=L_l, Ssurf=S_{LF} and J=$j_{F,S}$.

The specification of BM2 does not give any information how to choose the concentration boundary layer thickness. Therefore, in order to verify the 1d concept for irregular biofilms under different flow conditions, a parameter fitting study is carried out, trying to determine the new model parameter L_l from the 3d simulation results. Since this requires 3d data, such a parameter-fitting approach cannot be considered a predictive solution of BM2 itself. It is included here in order to investigate how good the correlation of L_l and the hydrodynamic conditions in the reactor segment can be expected to be.

The boundary-layer thickness can be estimated from 3d data in several ways. Not only can one use different methods to solve the underlying 1d differential equations, *e.g.*, the methods described in BM1, but also different criteria can be formulated to judge how well 3d and 1d data match. Ideally, one simultaneously wants to minimize differences for $j_{F,S}$ and S_{LF},

$$\min_{L_L} \left(j_{F,S,3d} - j_{F,S} \right)^2, \quad \min_{L_L} \left(S_{LF,3d} - S_{LF} \right)^2 \qquad (4.10)$$

where the superscripts denote the values that were obtained from the 3d simulation. However, matching both quantities at the same time typically cannot be accomplished. Multi-criteria problems of the type have no unique optimality criterion. Different multi-criteria concepts have different advantages and drawbacks that must judged from case to case (e.g., Stadler, 1988).

A multi-criteria optimization concept appropriate for this kind of problems is the *Edgeworth-Pareto* optimality concept. It states that a specific choice of value L_l is an Edgeworth-Pareto optimal compromise if further improvement of S_{LF} is only possible at the expense of a worse $j_{F,S}$ or vice versa. However, also this Edgeworth-Pareto compromise is not unique. In general, infinite choices of parameters can satisfy this compromise, and they cannot be distinguished objectively on the basis of neutral mathematical arguments; thus, one is considered as good as the other. It is, therefore, up to the modeler to make this choice.

Typically this is done *a priori* and often implicit by transforming the multi-criteria optimization problem into a regular scalar optimization problem that has a unique solution, which is Edgeworth-Pareto optimal for the original multi-criteria problem ("scalarization"). Many strategies can achieve this (Göpfert and Nehse, 1990). If the concentration boundary layer concept is physically meaningful, we would expect, however, that different physically reasonable choices lead to similar parameter values for L_l. Otherwise, L_l is an arbitrary fitting parameter.

This study includes the following choices of a compromise, not all of which necessarily lead to a strict Edgeworth-Pareto optimum:

- minimizing the sum of deviations between model solution and reference data: flux of substrate across the biofilm surface, $j_{F,S}$, average concentration of the substrate at the biofilm surface, S_{LF}, as well as the average substrate concentration at the substratum, S_0, and the biomass production rate (N1)
- using (N1b) $j_{F,S}$, (N1c) S_{LF}, and (N1d) S_0 (the average concentration at the substratum) as sole criterion for adjustment of the concentration boundary layer thickness.
- minimizing the sum of the relative deviations in S_{LF} and $j_{F,S}$ (PA)
- minimizing the error in $j_{F,S}$, postulating the relative deviations of S_{LF} and $j_{F,S}$ to be equal and enforcing an *ad hoc* global mass balance on the 1d-reactor under completely mixed conditions as in BM1 (see Morgenroth 2003) (N1g)

Not all of these approaches are carried through for all cases. In particular, N1b-d were only applied to selected cases for the purpose of comparison. The performance of the various parameter estimation procedures is discussed here (see also Table 4.14).

Comparing the various fitting approaches to BM2 shows that the values obtained for L_l agree within reasonable bounds. The differences in the results can be attributed to the different mathematical methods to solve the 1d steady-state biofilm model (as discussed for BM1) and to the choice of the fitting criteria that were used. However, as shown in the L_l-L_l-diagram in Figure 4.10, in the range of L_l values obtained in BM2, uncertainties of 10% or more are not well dampened when these data are used as input parameters to calculate $j_{F,S}$ and S_{LF}. Therefore, generally one must be very careful in determining the additional model parameter L_l.

Table 4.14. Concentration boundary layer thickness [μm] obtained by parameter fitting procedures with 1d models for the wavy biofilm architecture

Model	W1	W2	W3	W4	W5	W6	W7
N1	75	94	75	94	75	94	75
N1b	73.1	132	72.6	133	81.7	141	N/A
N1c	73.7	135	74.0	136	70.8	129	74.9
N1d	N/A	124	N/A	125	68.9	125	69.76
PA	73.0	93	74	95	71	83	74
A	73.2	109	73.8	116	77.4	88.6	91.3

All simulations show that the boundary-layer thickness depends more on the hydrodynamic situation than on the thickness of the biofilm (and, hence, the amount of substrate converted in the biofilm or the microbial activity). It is noticed that the fitted value of L_l clearly depends on hydrodynamics by comparing the results of Cases W1, W3, W5, W7 (fast flow) with the ones of W2, W4, W6 (slow flow). Other variables, such as the bulk substrate concentration and the biofilm thickness, play no or only minor roles in these experiments. The exception to this conclusion is N1g, where the values of L_l fluctuate more.

In this approach, stronger conditions were imposed on the minimization problem than in all the other approaches. Generally, the fitted value decreases with increasing free-flow velocity, as was seen in the 3d simulations (Figure 4.10). However, carrying over the value L_L found for one biofilm structure to another one under the same hydrodynamic condition is not possible.

In order to be able to solve BM2 by a 1d model, a correlation between L_L and the data provided in the problem specification must be found. We shall discuss two different approaches.

4.3.3.3.2 Numerical steady state model based on a global mass balance: N1s

Approach N1s is derived from a steady state mass balance as introduced in Section 3.5. The concentration boundary layer thickness L_L is eliminated from the problem description by incorporating the boundary conditions (4.1). This concept explicitly uses the hydrodynamic information provided in the problem description and, thus, links L_L to the flow velocity in the reactor segment, as assumed in Section 4.3.3.2.2 ($u(z)$ from equation (4.6), see Figure 4.8).

The substrate concentration is kept at a constant value S_B in the influent. Hence, the amount of substrate transported into the system is given by

$$F_{in,S} = Q\,S_B = S_B L_Y \int_0^{L_z} u(z)\,dz \qquad (4.11)$$

By assuming $\partial S/\partial x=0$ in the boundary layer, the outflow flux is

$$F_{ef,S} = L_Y \int_0^{L_z} u(z)\,S(z)\,dz \qquad (4.12)$$

One considers a mass transfer boundary layer of a priori *unknown* thickness L_L and a linear concentration gradient perpendicular to the substratum, so that:

$$S(z) = S_{LF} + (z - L_F)(S_B - S_{LF})/L_L \qquad \text{if } L_F < z < L_F + L_L \qquad (4.13)$$
$$S(z) = S_B \qquad \text{if } z > L_F + L_L$$

where the concentration S_{LF} at the biofilm-liquid interface is unknown at this point.

After integration of (4.11) and (4.12), the substrate consumed in the biofilm is

$$F_{F,S} = F_{in,S} - F_{ef,S} = -\frac{2}{3}L_L^2\left(S_B - S_{LF}\right)\frac{u_{X,max}L_Y}{L_z - L_F} = -D_F A_F \frac{dS}{dz}\bigg|_{LF} \qquad (4.14)$$

L_L can be eliminated from the flux continuity condition across the biofilm/liquid interface:

$$D_F \frac{dS}{dz}\bigg|_{LF} = D_L \frac{S_B - S_{LF}}{L_L} \qquad (4.15)$$

Finally, one obtains the nonlinear equation for S_{LF}

$$\frac{2}{3}\frac{L_Y u_{X,max}\left(D_F/D_L\right)^2}{L_z - L_F}\frac{\left(S_B - S_{LF}\right)^3}{\left(dS/dz|_{LF}\right)^2} - D_F A_F \frac{dS}{dz}\bigg|_{LF} = 0 \qquad (4.16)$$

As in section 4.3.3.3, a numerical solution of (3.21) is needed for solving equation (4.16).

4.3.3.3.3 General Mass Transfer Correlation: N1sc

An alternative one-dimensional approach to estimate L_L from input data is provided by N1sc. The contribution of hydrodynamics to the mass transfer is taken into account by explicit general empirical correlations connecting both effects, although these correlations originally have been developed for systems other than biofilms. This approach to BM2 is based on the observation that mass transfer depends more heavily on the hydrodynamics than on the surface heterogeneity of the biofilm. In particular, the mass flux $j_{F,S}$ into the biofilm is estimated using the correlation of Comiti *et al.* (2000) in terms of the Sherwood number (Sh). This correlation estimates Sh for a range of regular systems, e.g., packed beds of spheres, cylinders, plates for linear and non-linear laminar flow, as well as in smooth tubes for turbulent conditions ($0.03 < Re < 10^5$). The generalized correlation is based on the wall energetic criterion Xe_w.

$$Sh = 3.66 + \alpha\, Xe^{11/48}\, Sc^{1/3} \tag{4.17}$$

where α depends on the specific porous media characteristics and also on the wall energetic criterion and *Sc* is the Schmidt dimensionless group (ratio of fluid kinematic viscosity and diffusion coefficient). Xe_w can be calculated using the following equation (Comiti *et al.* 2000),

$$Xe = 64 \cdot Re^2 \cdot (1 + 0.0121 \cdot Re) \tag{4.18}$$

Sh numbers computed for chemical reactors and for ducts are 15% higher than those computed for biofilms (Sh_b) using the same hydrodynamic conditions (Nicollela *et al.* 2000). Introducing this correction factor for biofilm systems and assuming tortuosity values around 1 lead to

$$Sh_b \approx 0.85 \cdot Sh = 0.85 \cdot \left(3.66 + 0.101 \cdot Xe_w^{11/48} \cdot Sc^{1/3}\right) \tag{4.19}$$

Once the Sherwood number is estimated for the biofilm system, the concentration boundary layer thickness can be computed with the expression,

$$L_L = \frac{L_{L,max}}{Sh_b} \tag{4.20}$$

where $L_{L,max}$ is the maximum depth of the liquid region (Figure 4.6). Using this as an input data for the 1d steady-state biofilm model, the surface concentration S_{LF} can be calculated by numerical methods for two-point boundary value problems. The results reported here are computed by a high accuracy orthogonal collocation method (Finlayson 1972) on a grid of 14 nodes in the biofilm.

4.3.4 Results and discussion

4.3.4.1 System behavior as revealed by 3d simulation

The numerical simulations carried out with N3c for BM2 show that the substrate concentration at the liquid/biofilm interface decreases as the biofilm becomes thicker, due to increased consumption in the biofilm. Since the biofilm thickness does not alter the inflow boundary conditions for the substrate, the total amount of substrate transported into the channel system does not change with the base film thickness, being $S_B \int_{A_{in}} u\, dydz$, i.e., the flow rate multiplied with the bulk substrate concentration. Hence, a decreased surface concentration implies an increased flux $j_{F,S}$. This is confirmed by the numerical values that are summarised in Tables 4.15a and 4.15b.

Table 4.15a. Results for the flat biofilm architecture. Reported are average surface concentration $S_{LF,avg}$ [g/m³] and total mass flux $j_{F,S}$ [10^{-7}g/d] into the biofilm. The last three rows contain selected values obtained by different parameter fitting approaches

Model	Case F1		Case F2		Case F3		Case F4	
	$S_{LF,avg}$	$j_{F,S}$	$S_{LF,avg}$	$j_{F,S}$	$S_{LF,avg}$	$j_{F,S}$	$S_{LF,avg}$	$j_{F,S}$
N3c	3.82	1.32	3.20	1.99	2.86	1.10	2.17	1.43
N2b	3.83	1.32	3.20	1.94	2.87	1.09	2.18	1.40
N2d	3.99	1.25	3.19	1.96	2.94	1.04	2.16	1.40
N1s	3.83	1.32	3.23	2.01	2.89	1.11	2.22	1.46
N1sc	3.76	1.30	3.15	1.94	3.08	1.15	2.42	1.55

N1	3.80	1.32	3.21	1.96	2.85	1.10	2.21	1.42
PA	3.81	1.32	3.20	2.00	2.85	1.10	2.16	1.43
A	3.61	1.39	3.40	1.86	2.66	1.18	2.40	1.28

Table 4.15b. Results for the wavy biofilm architecture. Reported are average surface concentration $S_{LF,avg}$ [g/m³] and total mass flux $j_{F,S}$ [10^{-7}g/d] into the biofilm. The last three rows contain selected values obtained by different parameter fitting approaches.

Model	Case W1		Case W2		Case W3		Case W4	
	$S_{LF,avg}$	$j_{F,S}$	$S_{LF,avg}$	$j_{F,S}$	$S_{LF,avg}$	$j_{F,S}$	$S_{LF,avg}$	$j_{F,S}$
N3c	2.50	4.59	1.48	2.95	2.41	4.76	1.42	2.99
N2b	2.35	4.38	1.41	2.81	2.26	4.49	1.36	2.84
N2d	3.02	5.21	1.38	2.51	2.89	5.48	1.27	2.61
N2e	3.03	5.14	1.36	2.52	2.90	5.44	1.29	2.59
N1s	2.61	4.72	1.60	3.12	2.52	4.89	1.54	3.17
N1sc	2.71	4.83	1.58	3.07	2.63	4.99	1.54	3.11
N1a	N/A	3.34	N/A	3.34	N/A	3.41	N/A	3.41

N1	2.50	4.60	1.50	2.95	2.43	4.74	1.45	2.98
PA	2.51	4.58	1.49	2.92	2.41	4.71	1.42	2.94
A	2.66	4.29	1.69	2.54	2.65	4.29	1.65	2.50

Model	Case W5		Case W6		Case W7	
	$S_{LF,avg}$	$j_{F,S}$	$S_{LF,avg}$	$j_{F,S}$	$S_{LF,avg}$	$j_{F,S}$
N3c	3.59	2.61	2.54	2.10	0.226	0.502
N2b	3.54	2.65	2.41	2.10	0.21	0.47
N2d	3.82	2.53	2.34	1.85	0.27	0.610
N2e	3.88	2.51	2.42	1.81	0.26	0.598
N1s	3.64	2.69	2.60	2.20	0.24	0.518
N1sc	3.71	2.71	2.58	2.17	0.24	0.529
N1a	N/A	2.10	N/A	2.10	N/A	0.35

N1	3.56	2.61	2.47	2.13	0.229	0.502
PA	3.59	2.67	2.54	2.17	0.23	0.496
A	3.45	2.71	2.41	2.21	0.32	0.276

Table 4.15c. Results for the bumpy biofilm architecture. Reported are average surface concentration $S_{LF,avg}$ [g/m³] and total mass flux $j_{F,S}$ [10^{-7}g/d] into the biofilm. The last three rows contain selected values obtained by different parameter fitting approaches.

Model	Case B1		Case B2		Case B3		Case B4	
	$S_{LF,avg}$	$j_{F,S}$	$S_{LF,avg}$	$j_{F,S}$	$S_{LF,avg}$	$j_{F,S}$	$S_{LF,avg}$	$j_{F,S}$
N3c	2.59	3.13	1.19	5.02	3.35	3.63	1.78	7.24
N2b	2.41	3.33	1.08	4.92	3.14	3.92		6.87
N2d	2.01	2.66	0.95	4.04	3.02	3.64	1.91	7.77
N2e	2.29	2.63	0.95	4.03	3.39	3.51	1.91	7.68
N1s	2.67	3.50	1.29	5.57	3.42	4.05	1.91	7.93
N1sc	2.67	3.44	1.29	5.48	3.60	4.09	2.09	8.48
N1	2.57	3.41	1.20	5.06	3.34	4.00	1.78	7.30
PA	2.59	3.42	1.19	5.15	3.34	3.99	1.78	7.42
A	2.42	3.34	1.39	4.17	3.12	3.88	1.99	6.39

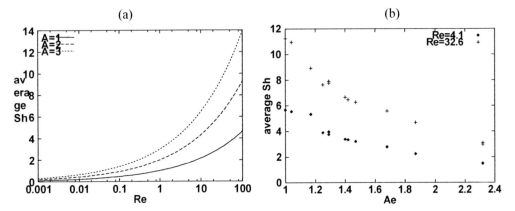

Figure 4.11. (a) Relationship between average Sherwood number (measuring mass transfer) and Reynolds number (measuring flow velocity) as found in Eberl et al (2000a) for biofilm systems with Dirichlet inflow conditions, Sh = A Re$^{1/3}$. The parameter A depends on the biofilm geometry and on the Schmidt number Sc, i.e. the ratio of fluid viscosity v and diffusion coefficient D (one has Pe= Sc Re). (b) Relationship between average Sherwood number Sh (measuring mass transfer) and Area enlargement number A_e (measuring surface irregularity) for various wavy biofilms under two different bulk flow velocities, as reported in Eberl et al (2000a) for biofilm systems with Dirichlet inflow conditions.

The three-dimensional simulations for BM2 are complemented by a comprehensive study of the variation of the total flux of substrate into the biofilm as it depends flow velocity and biofilm surface roughness. As shown in Figure 4.11, $j_{F,S}$ increases as the flow velocity increases. More specifically, Sh= A Re$^{1/3}$, where Re is the Reynolds number describing the flow conditions and Sh is the average Sherwood number. The constant A depends on the biofilm geometry and the Schmidt Sc number, which relates fluid viscosity and the diffusion coefficient D. Furthermore, if the hydrodynamic load is kept the same, Sh decreases as A_e increases for wavy biofilm architectures.

4.3.4.2 Comparison of models in BM2 and their performance

The results obtained for BM2 are presented in Tables 4.15a-c. We use the average surface concentration of the dissolved substrate, $S_{LF,avg}$ and the flux of dissolved substrate from the liquid region into the solid biofilm region, $j_{F,S}$, as indicators for model performance, both at steady state. Due to the sheer amount of computations carried out, the results are subdivided.

We organise them by types of surface biofilm layer, *i.e.* we present them separately for the flat, wavy and bumpy structures. The data presented comprise the predictive models N3c, N2b, N2d, N2e, N1s, N1sc. The results of N1a are included only in one figure, for the purpose of illustration, since this approach by construction is not sensitive to hydrodynamics.

In a quantitative comparison of $S_{LF,avg}$ and $j_{F,S}$ obtained by different models, one must be aware that they are affected by two main factors: (i) the modelling approach *per se* and (ii) the numerical methods used for the model solution and for the *a posteriori* calculation of $S_{LF,avg}$ and $j_{F,S}$ from the simulation results (where applicable). Numerical methods for nonlinear and for large-scale linear problems involve iterative solvers, which must be terminated once an appropriate accuracy is reached. Choosing the parameters of the numerical method that control accuracy, e.g. the stopping criteria for iterations, refinement of dicretisation, however, is a trade-off with computing speed. Making this choice and putting emphasis on computational effort or on accuracy (given the uncertainties in the input data) is an inherent part of model application. Therefore, one cannot expect a total agreement between different model approaches. Although only (i) should be investigated here, the effects of (ii) sometimes cannot be neglected. The results of Morgenroth *et al* (2004) might serve as a guide to characterize and judge the numerical uncertainties. A further indication can be derived from the results for the flat biofilm geometries, where N3c, N2b, N2d, N2e use the same data and solve identical problems Moreover, N3c, uses the analytical solution for the flow field, which rules out numerical errors in these cases. In the following discussion the numerical solutions obtained by the complete three-dimensional model description are used as reference data for the other models. Also here one must be aware that these are numerical approximations for a complex flow and mass transfer problem, and therefore subject to numerical inaccuracies themselves.

4.3.4.2.1 Qualitative system behavior

A first test for the simplified models is whether or not they render the same macro-scale behavior as the full 3d model with hydrodynamics. Tables 4.15 confirm that the simplified models achieve similar behavior for all cases of BM2.

For the homogeneous biofilm, we find that the average surface concentration $S_{LF,avg}$ is highest in Case F1 (thin biofilm, fast flow) and smallest in Case F4 (thick biofilm, slow flow); more specific we have the ordering Case F1 > Case F2> Case F3 > Case F4. For the flux $j_{F,S}$, on the other hand, we obtain Case F2 > Case F4> Case F1> Case F3.

For the bumpy biofilm architecture, we find for the average surface concentration $S_{LF,avg}$ that Case B3 > Case B1 > Case B4 > Case B2. The flux $j_{F,S}$ is ordered as Case B4 > Case B2 > Case B3 > Case B1. These strict orderings are obtained because the values in the different cases are nicely separated from each other. Accordingly, all predictive models reflect this ordering.

The trends differ for the wavy biofilm structure, where the results for Cases 2 and 4 are very similar. While most models find that $S_{LF,avg}$ behaves like Case W5 > Case W6 > Case W1> Case W3 > Case W2 > Case W4 > Case W7 and $j_{F,S}$ like Case W3 > Case W1> Case W4 > Case W2 > Case W5 > Case W6 > Case W7, this is not true for all approaches, in particular for the approaches N2d and N2e, in the cases of a fast flowing bulk liquid. This might be due to the simplification of the hydrodynamic description.

In summary, however, we can state that all models show approximately the same qualitative trends and sensitivity to variations in both investigated parameters of physical complexity.

4.3.4.2.2 Quantitative comparison of 2d models N2b, N2d and N2e

A comparison of the simulation results in Tables 4.15a,b,c shows that, in general, N3c and N2b agree, albeit with quantitative discrepancies. Comparing the results of N2b with the results of N2d shows that giving up the conservation principles for the fluid flow in the domain leads to a further increased deviation with reference to N3c. The magnitude of this change is in some cases small (<2%, typically at slow bulk flow) and in other cases considerable (>20%, typically at fast bulk flow). Therefore, an *a priori* assessment of the effect of the simplification of the hydrodynamics in a particular case cannot be given without further analysis of the hydrodynamic conditions. On the other hand, comparing the results of N2d and N2e shows that both approaches to reduce the dimensionality of the problem lead to similar results, at least in the case of a moderately heterogeneous biofilm architecture. This leads to the conclusion that, from a global (macro-scale) point of view, model reduction is much more sensitive to calculation of the hydrodynamic flow field than to its geometrical description (under the hypothesis that biomass and volume are maintained).

4.3.4.2.3 Quantitative comparison of closed 1d models N1s and N1sc

Both 1d approaches consider hydrodynamics, but do not explicitly consider information about the geometrical structure of the biofilm architecture other than the average biofilm thickness or amount of biomass. At the core of both models is a 1d biofilm model that is solved numerically. The results presented above show that these one-dimensional models can give results that are as accurate as the 2d simulations of N2b, sometimes even better when the comparison is to the 3d output.

For the wavy biofilms geometries, N1s seems to be more accurate than N1sc for a higher flow velocity (Cases W1, W3, W5, W7) and *vice versa* for slower hydrodynamics (Cases W2, W4, W6). This impression is emphasized more in the case of the bumpy biofilm geometries. The situation is different with the flat biofilm geometry, where N1s seems to perform better than N1sc in all cases. This is not surprising, since in deriving N1s, the heterogeneous biofilm structure is approximated by a flat one, while the empirical correlation, upon which N1sc is based, is not developed for this idealized system, but explicitly considers the surface irregularity, albeit in an across-the-board manner.

The roughness of a biofilm architecture is a consequence of the hydrodynamic conditions in a biofilm (Van Loosdrecht *et al.* 1997; Picioreanu *et al.* 2000b): a faster flow induces a smoother biofilm due to increased nutrient availability and stronger detachment, while a slower flow field induces are more heterogeneous structure. Under this light, N1s gives good results on one end of the spectrum (fast and smooth), while N1sc is the preferred method of choice at the other end (slow and rough). In cases that cannot be categorized uniquely as belonging to one or the other group, a combination of both models in the sense of averaging can be of advantage.

4.3.4.2.4 Comparing parameter fitting approaches with predictive 1D models

The parameter-fitting approaches (N1, N1b-d, PA) allow a better approximation of the 3d results of N3c for the non-flat biofilm than the closed one-dimensional models N1s and N1sc. As parameter-fitting solutions, they implicitly take the spatial heterogeneity of the biofilm into account, which is inherent in the reference data that were used for calibration. On the other hand, when deriving N1s, the biofilm geometry is approximated by a 1d flat biofilm. Not surprisingly, therefore, N1s reaches the accuracy of N1 and N1b-d only in the case of the flat biofilm. On the other extreme is N1sc, which performs with the least accuracy in the flat case, as already discussed above.

Note that for every situation at least one of N1s or N1sc gives satisfactory results (depending on the hydrodynamic condition and the biofilm architecture). Since it can be predicted which of these is likely to deliver better results (depending on flow and biofilm roughness), a 1d predictive model can be found. Nevertheless, the parameter-fitting approaches (N1, N1b-d, PA) indicate that it might even be possible to achieve improvements if the special biofilm structure can be taken in to account in determining the concentration boundary layer. How this correlation could look like is currently an open problem in biofilm modeling. Note that the parameter fitting approach (A) does not confirm these findings. This is due to the heavy restriction that was posed as a constraint of the minimization problem.

4.3.4.3 Comparison of Model Requirements

A further distinction between the models in this survey must be made based on the input data that are necessary. While kinetic or physical model parameters are the same, the situation is quite different regarding the biofilm description. The supposedly higher accuracy of the 2d and 3d models (N3c, N2b, N2d and N2e) comes at the high cost of requiring a detailed description of the biofilm architecture on the meso-scale, *i.e.* on a length scale of around 100 µm to 1 mm. This scale can be investigated experimentally only with modern microscopy and microelectrode techniques. For many applications, investigation at this small scale is not feasible and often not of interest. On the other hand, the predictive 1d models require only lumped descriptions, such as average height of a biofilm or total biomass.

Even if an accurate description of the biofilm architecture is at hand, the 3d simulation raises the issue of computational effort, in particular regarding the calculation of the flow field. The reduction to a 2d description relaxes this concern. The further reduction to a 1d description leads to biofilm models that can be implemented easily within off-the-shelve software packages (see Section 3.4.3 and 3.5.3). Furthermore, these 1d models can be embedded in reactor models that allow for the calculation of the biofilm characteristics by reactor balances.

4.3.4.4 Lessons learned from BM2

The numerical experiments carried out for BM2 investigate the question of whether a detailed 3d representation of local biofilm processes is always necessary in order to obtain global lumped results about mass transfer in a biofilm reactor segment. The 1d, 2d, and 3d models considered showed the same general sensitivities towards changes in biofilm thickness and hydrodynamics and were all able to describe the qualitative system behavior. For the quantitative details, the key to a successful model reduction is a good description of the hydrodynamic conditions in the reactor segment. In a 2d reduction, this can be accomplished by a 2d version of the governing flow equations. The 1d approaches require a global mass balance or an empirical correlation that incorporates the hydrodynamics with passable reliability and accuracy.

Which simplified predictive models offers the best effort/accuracy/reliability trade-off depends largely on the hydrodynamic regime. Therefore, an analysis of the flow conditions in the reactor is required first before a simplification should be applied. If one aims for a 1d description, N1s is the preferred method for fast flow regimes, while N1sc is the method of choice for slower flow velocities. N2b is a safe choice for 2d descriptions in all flow situations, but N2d and N2e offer a less computing-expensive alternative under slower bulk flow. Due to the enormous requirement of input data, the application of the 3d hydrodynamic model N3c is restricted.

Finally, one must be aware that these statements are made for applications in which only global results are of interest; that is, no refined resolution of the processes inside the biofilm is required. If such local results are desired, 1d models cannot yield a good description for spatially heterogeneous biofilms. At least a 2d model must be applied and, hence, the required input data and computing power must be provided.

4.4   BENCHMARK 3: MICROBIAL COMPETITION

The goal of multi-species benchmark problem 3 (BM3) is to evaluate the ability of the different biofilm models to describe microbiological competition. In particular, BM3 focuses on competition for the same substrate and the same space in a biofilm. Each type of competition is sufficiently studied so that it can be represented mathematically. Together, these competitions provide a rigorous test for modeling multi-species biofilms, but without introducing unnecessary complexity. To meet the goal, BM3 includes three biomass types having distinctly different metabolic functions:
  • aerobic heterotrophs
  • aerobic, autotrophic nitrifiers
  • inert (or inactive) biomass
   This scenario represents a common situation for biofilms in nature and in treatment processes for wastewater and drinking water. Situations similar to the benchmark have been modeled in different ways by many researchers over the years (e.g., Kissel *et al.* 1984; Wanner and Gujer, 1986; Furumai and Rittmann 1992, 1994; Rittmann and Manem, 1992; Reichert, 1994; Wanner and Reichert, 1996; Rittmann and Stilwell, 2002).

4.4.1 Definition of the system modeled

For simplicity and comparability, the multi-species benchmark uses the same physical domain as the standard case of BM1: a flat biofilm substratum in contact with a completely mixed reactor experiencing a steady flow rate. The standard parameters defining the physical domain are listed in Table 4.1. To avoid unnecessary complexity, BM3 treats the nitrifiers as one "species" that oxidizes NH_4^+-N directly to NO_3^--N. Thus, it does not consider the intermediate NO_2^- or the division of nitrifiers between ammonia oxidizers and nitrite oxidizers. Active heterotrophs and nitrifiers follow Monod kinetics for substrate utilization and synthesis. They also undergo decay following two paths: (1) lysis and oxidation by endogenous respiration, and (2) inactivation to form inert or inactive biomass. The inert biomass does not consume substrate, and it is not consumed by any reactions. All forms of biomass can be lost by physical detachment.
   Table 4.16 summarizes the stoichiometry of all the reactions involving the three biomass types -- X_H for heterotrophs, X_N for nitrifiers, and X_I for inerts -- and three substrates -- S_S for the organic donor substrate (as COD), S_N for NH_4^+-N, and S_{O2} for dissolved oxygen. For comparability, the multi-species benchmark used the same Monod kinetics and stoichiometry for the heterotrophic bacteria as in the first benchmark problem. Table 4.2 defines the heterotrophic biomass parameters and their values. Table 4.17 lists the comparable parameters for the nitrifying bacteria. In order to avoid making BM3 more complicated than necessary to encompass microbiological competitions, the Task Group included neither uptake of NH_4^+-N for synthesis nor its release during respiration.

4.4.2 Cases investigated

The standard case, represented by the physical conditions of Table 4.1 and the reaction conditions of Tables 4.16 and 4.17, provide a rigorous comparison among the different models, because it allows all biomass types to be present in significant amounts and also gives significant substrate gradients in the biofilm. In addition to the standard case, the Task Group altered physical or reaction conditions in order to create five special cases that emphasize certain aspects of a multi-species biofilm. Table 4.18 summarizes the standard case and the five special cases, including their conditions and purposes.

Table 4.16. Matrix of stoichiometry and kinetic expressions for multi-species BM3

Process name	Biomass Types			Substrates			Kinetic Expressions
	X_H	X_N	X_I	S_S	S_N	S_{O2}	
Heterotroph metabolism	1			$\dfrac{-1}{Y_H}$		$\dfrac{-(1-Y_H)}{Y_H}$	$\mu_{max,H} X_H \dfrac{S_S}{K_S+S_S} \dfrac{S_{O2}}{K_{O2,H}+S_{O2}}$
Heterotroph inactivation	-1		1				$b_{ina,H} X_H$
Heterotroph respiration	-1					-1	$b_{res,H} X_H \dfrac{S_{O2}}{K_{O2,H}+S_{O2}}$
Autotroph metabolism		1			$\dfrac{-1}{Y_N}$	$\dfrac{-(4.57-Y_N)}{Y_N}$	$\mu_{max,N} X_N \dfrac{S_N}{K_N+S_N} \dfrac{S_{O2}}{K_{O2,N}+S_{O2}}$
Autotroph inactivation		-1	1				$b_{ina,N} X_N$
Autotroph respiration		-1				-1	$b_{res,N} X_N \dfrac{S_{O2}}{K_{O2,N}+S_{O2}}$

Table 4.17. Kinetic and stoichiometric parameters for the nitrifiers in BM3

Parameter	Symbol	Value with Units
Maximum specific utilization rate	$q_{max,N} = (\mu_{max,N}/Y_N)$	2.2 $g_{COD-N}/g_{COD-X}d$
Substrate half-maximum-rate concentration	K_N	1.5 g_N/m^3
True yield	Y_N	0.063 g_{COD-X}/g_{COD-N}
Decay/respiration rate coefficient	$b_{res,N}$	0.12 d^{-1}
Decay/inactivation rate coefficient	$b_{ina,N}$	0.03 d^{-1}
Diffusion coefficient in pure water	D_N	1.7×10^{-4} m^2/d
Ratio of diffusion coefficient in biofilm versus water	$D_{F,N}/D_N$	1
Oxygen half-maximum-rate concentration	$K_{O2,N}$	0.5 g_{O2}/m^3
Diffusion coefficient of O_2 in pure water	D_{O2}	2×10^{-4} m^2/d
Ratio of diffusion coefficient in biofilm versus water	$D_{F,O2}/D_{O2}$	1

4.4.3 One-dimensional models applied

Nine 1d models could solve BM3, although they differ in their degree of complexity and user accessibility. The common and distinguishing characteristics of the nine models are described in two steps. The first step summarizes a general model that captures all phenomena in the multi-species benchmark problem in a mathematical framework that does not involve making additional simplifications. Two of the 1d models are of this general type. The second step identifies simplifying assumptions used in the other seven 1d models.

Table 4.18. Summary of the standard and five special cases used in BM3

Case Name	Condition	Purpose
Standard	• Given by Tables 4.1, 4.16 and 4.17	• Have each biomass type be a significant fraction in the biofilm • Have significant concentration gradients in the biofilm
High N:COD ratio	• Increase $S_{in,N}$ from 6 to 30 g_N/m^3	• Make nitrifiers a dominant factor in the biofilm • Emphasize dilution by nitrifiers
Low N:COD ratio	• Decrease $S_{in,N}$ from 6 to 1.2 g_N/m^3	• Make nitrifiers a minor player that is affected strongly by the heterotrophs • Emphasize role of protection of slow-growing species
High production of inerts	• Increase $b_{ina,H}$ to 0.2/d, while decreasing $b_{res,H}$ to 0.2/d • Increase $b_{ina,N}$ to 0.075/d, while decreasing $b_{res,N}$ to 0.075/d	• Emphasize the role of inerts in diluting heterotrophs
High detachment to give a thin biofilm	• Increase the detachment rate so that the steady-state biofilm thickness is reduced from 500 µm to 20 µm	• De-emphasize the role of dilution by inerts or nitrifiers
High oxygen sensitivity by nitrifiers	• Increase $K_{O2,N}$ from 0.5 to 5.0 g_{O2}/m^3	• Emphasize role of protection of slow-growing species

4.4.3.1 The general one-dimensional, multi-species, and multi-substrate model

Model N1 is of the general type and follow the mathematical framework in this section without simplifications. Model N1 exploited the AQUASIM software (Reichert, 1994; Wanner and Reichert, 1996) and required numerical solution of the non-steady-state differential equations shown below. Many of the features of this general model carry over to the other seven models, although they make a range of simplifications to reduce the computational burden. Therefore, a summary of model N1 provides an excellent foundation for understanding all the models.

The steady-state differential mass balance on any substrate, along with its boundary conditions, takes the form

$$D_{F,i}\frac{d^2 S_{F,i}}{dz^2} + r_{S,i} = 0 \qquad (4.21)$$

$$\frac{dS_{F,i}}{dz} = 0 \qquad \text{at } z = 0 \qquad (4.22)$$

$$S_{F,i} = S_{LF,i} \qquad \text{at } z = L_f \qquad (4.23)$$

where $D_{F,i}$ = diffusion coefficient for a substrate in the biofilm (L^2T^{-1}), $S_{F,i}$ = concentration of the substrate at a position in the biofilm ($M_S L^{-3}$), z = coordinate normal to the biofilm surface (L), $S_{LF,i}$ = concentration of the substrate at the biofilm/liquid-film interface ($M_S L^{-3}$), $r_{S,i}$ = transformation rate of the substrate ($M_S L^{-3}T^{-1}$), and L_F = biofilm thickness (L).

The differential mass balance on any biomass type is

$$0 = -\frac{d(u_F X_{F,i})}{dz} + r_{X,i} \qquad (4.24)$$

$$u_F X_{F,i} = 0 \qquad \text{at} \qquad z = 0 \tag{4.25}$$

$$X_{F,Tot} = X_{F,H} + X_{F,N} + X_{F,I} = \sum_i X_{F,i} \tag{4.26}$$

where $X_{F,i}$ = concentration of a biomass type ($M_X L^{-3}$), u_F = velocity at which biomass moves away from the substratum (LT^{-1}), and $r_{X,i}$ = transformation rate of the biomass type ($M_X L^{-3} T^{-1}$). Equations (4.24) and (4.26) allow the distribution of the three biomass types to change with location, while requiring that the total biomass density ($X_{F,Tot}$) at any location in the biofilm be constant.

The rate of transformation for any substrate (i.e., COD, NH_4^+-N, or O_2) and any biomass component (i.e., heterotrophs, nitrifiers, or inerts) used in a mass balance equation is calculated using the stoichiometric factors and kinetic expressions given in Table 4.16:

$$r_{S,i} = \sum_j v_{si,j} \rho_j \tag{4.27}$$

$$r_{X,i} = \sum_j v_{xi,j} \rho_j \tag{4.28}$$

where $v_{si,j}$ = stoichiometric factor associated with substrate i and transformation process j ($M_X M_S^{-1}$), $v_{xi,j}$ = stoichiometric factor associated with biomass type i and transformation process j (non dimensional), and ρ_j = kinetic expression of transformation process ($M_X L^{-3} T^{-1}$) in Table 4.16.

The biomass displacement velocity u_F is computed from

$$u_F(z) = \frac{1}{X_{F,Tot}} \int_0^z \sum_i r_{X,i} dz \tag{4.29}$$

where $X_{F,Tot}$ = total biomass concentration of the biofilm ($M_X L^{-3}$).

Mass transport through the liquid boundary layer is modeled as

$$j_{F,i} = \frac{D_i}{L_L} (S_{B,i} - S_{LF,i}) \tag{4.30}$$

where $j_{F,i}$ = flux of substrate i across the liquid boundary layer ($M_S L^{-2} T^{-1}$), D_i = diffusion coefficient in the bulk liquid ($L^2 T^{-1}$), L_L = thickness of the liquid boundary layer (L) , and $S_{B,i}$ = substrate bulk concentration ($M_S L^{-3}$).

The reactor's bulk-liquid volume is completely mixed, has steady input and output, and has no reactions other than substrate flux into the biofilm

$$0 = Q(S_{in,i} - S_{B,i}) + A_F j_{F,i} \tag{4.31}$$

where Q = flow rate ($L^3 T^{-1}$), $S_{in,i}$ = substrate influent concentration ($M_S L^{-3}$), A_F = total reactor area (L^2) (biofilm surface area = biofilm substratum area).

Finally, the change in biofilm thickness can be represented as depending on detachment that occurs solely at the outer surface (Equation (4.32)) or that is averaged over the entire biofilm volume (Equation (4.33)):

$$\frac{dL_F}{dt} = u_F(z = L_F) - u_{de} \tag{4.32}$$

$$\frac{dL_F}{dt} = u_F(z = L_F) - b_{de} L_F \tag{4.33}$$

where u_{de} = detachment velocity (LT^{-1}), and b_{de} = detachment coefficient (T^{-1}). When the biofilm thickness is specified -- as it is in this benchmark problem -- the values of u_{de} or b_{de} are adjusted until the steady-state thickness equals the desired value.

4.4.3.2 Simplifications and distinguishing features of the models

Table 4.19 summarizes the simplifications and distinguishing features for each of the nine models. Also listed are key primary references in which details of models' foundations can be found. The models also are described in Chapter 3.

Table 4.19. Summary of the simplifying assumptions, distinguishing features, and key primary references underlying the 1d models used for BM3

Model	Simplifications	Other distinguishing features	Key primary references
N1	None; followed general model based on AQUASIM	Surface detachment (Eq. (4.32))	Wanner and Gujer 1985; Reichert 1994; Wanner and Reichert 1996
N1a		Surface detachment (Eq. (4.32))	
N1b	Uniform biomass distribution (Eq. (4.24) relaxed)	Used $D_F' = D_F/\varepsilon_l$ to compensate for volume-based concentrations	
N1s		Volumetric detachment (Eq. (4.33))	NA
A	Uniform biomass distributions (Eq. (4.24) relaxed) Average of 1st-order and 0th-order analytical solutions for flux for each substrate $S_{LF:O2}$ was used instead of S_{O2} in Table 4.1	Volumetric detachment (Eq. (4.33))	Section 3.2
Aa	Uniform biomass distributions (Eq. (4.24) relaxed) 0th-order analytical solutions for flux of each substrate $S_{LF:O2}$ was used instead of S_{O2} in Table 4.1	Volumetric detachment (Eq. (4.33))	Section 3.2
PAa	Independent biomass distribution (Eqs. (4.24) and (4.26) relaxed) Concentrations of COD and NH$_3$-N were computed independently (Eq. (4.21) for O$_2$ relaxed) $S_{LF:O2}$ was used instead of S_{O2} in Table 4.1	Volumetric detachment (Eq. (4.33))	Rittmann and McCarty 1980; Sáez and Rittmann 1992; Rittmann and Stilwell 2002;
PAc	Nitrifiers and inerts existed in layers behind the heterotrophs (Eq. (4.24) relaxed) Concentrations of COD and NH$_3$-N were computed independently (Eq. (4.21) for O$_2$ relaxed) $S_{LF:O2}$ was used instead of S_{O2} in Table 4.1	Volumetric detachment (Eq. (4.33)) Detachment coefficient of nitrifiers and inerts was 1/10th that of heterotrophs Diffusion layer for nitrifiers increased by thickness of heterotrophic layer	
PAd	Uniform biomass distribution (Eq. (4.24) relaxed) Concentrations of COD and NH$_3$-N were computed independently (Eq. (4.21) for O$_2$ relaxed) $S_{LF:O2}$ was used instead of S_{O2} in Table 4.1	Volumetric detachment (Eq. (4.33))	

Note: Models N1, N1a, N1b, and N1s involve numerical solutions of the relevant differential equations; models PAa, PAc, PAd involve spreadsheet solutions based on pseudo-analytical solutions to relevant differential equations; and models A and Aa involve spreadsheet solutions to analytical solutions of the relevant differential equations.

The models can be categorized according to three characteristics: whether or not they require a numerical solution, how they distribute the biomass types throughout the biofilm, and how they handle oxygen limitation.

Models N1, N1a, N1b, and N1s require numerical solutions of the sets of differential equations and boundary conditions. N1, N1a, and N1b rely on the AQUASIM software, while N1s has its own software. Models PAa, PAc, and PAd exploit pseudo-analytical solutions for a steady-state biofilm (Rittmann and McCarty, 1980; Sáez and Rittmann, 1992). Models A and Aa exploit analytical solutions.

Using numerical models generally incurs significant computational demands, but -- at least in principle -- should give an accurate portrayal of the modeled phenomena. Models based on the pseudo-analytical solution, on the other hand, impose minimal computing demands and can be solved with a spreadsheet. The trade-off with pseudo-analytical solutions is that they require more simplifications to deal with the multi-species scenario. These simplifications might alter model outputs, compared to the numerical solutions. The analytical solutions (A and Aa) can be solved with a spreadsheet, but they require significant simplifications to represent multi-species phenomena. One key simplification is the choice of reaction order for the analytical solution. In Model A, the flux used was a weighted average of the analytical first-order flux ($j_{F,i}^{(1)}$) and the analytical zero-order flux ($j_{F,i}^{(0)}$):

$$j_{F,i} = \left(\frac{S_{LF,i}}{S_{LF,i} + K_i} \right) \cdot j_{F,i}^{(0)} + \left(1 - \frac{S_{LF,i}}{S_{LF,i} + K_i} \right) \cdot j_{F,i}^{(1)} \qquad (4.34)$$

The values of $j_{F,i}^{(0)}$ and $j_{F,i}^{(1)}$ can be computed from equations in Section 3.2. The weighting scheme in Eq. (4.34) makes a smooth transition from $j_{F,i}^{(0)}$ for $S_{LF,i} \gg K_i$ to $j_{F,i}^{(1)}$ for $S_{LF,i} \ll K_i$. Model Aa used only $j_{F,i}^{(0)}$.

The way in which a model represents the distribution of the three biomass types is the most important characteristic that distinguished them. Models N1 and N1a allow the distribution to develop naturally according to Equations (4.24), (4.26), and (4.32), with Equations (4.24) and (4.32) (i.e., surface detachment) being the keys for "moving" the biomass types differently in the biofilm. The other seven models relax Equation (4.24) and impose distributions. They also use an average volumetric detachment rate, Equation (4.33). Model PAa is the simplest case. The three types of biomass are independent of each other, which means that Equations (4.24)-(4.26) are relaxed (Section 3.3; Rittmann and Stilwell 2002). Each biomass type is present at $X_{F,Tot}$ and therefore, not diluted by any other type. Models N1b, N1s, PAd, A, and Aa impose uniform distributions throughout the biofilm; this relaxes Equation (4.24), but keeps Equation (4.26). The different types of biomass dilute each other evenly at all points in the biofilm. Model PAc imposes a layered distribution in which all heterotrophs are in the outer layer, all nitrifiers are in a middle layer, and all inerts are in a layer closest to the substratum (Section 3.3; Furumai and Rittmann 1992, 1994). This distribution is a simplified representation of the observed "fuzzy layering" of faster growing species on top of slower growing species (Kissel et al. 1984; Wanner and Gujer 1986; Rittmann and Manem 1992; Wanner and Reichert 1996). The inner layers are protected, which means that their b_{de} values are 10% of b_{de} for the heterotrophs. With layering, the different biomass types do not dilute each other, but the nitrifiers experience a reduced NH_4^+-N concentration because the NH_4^+-N diffuses through the heterotroph layer before reaching the nitrifier layer.

Oxygen limitation is another aspect that is treated differently among the models. Models N1, N1a, N1b, and N1s use the multiplicative-Monod kinetics shown in Table 4.16 for all positions in the biofilm, and this demands that they solve a mass balance on dissolved oxygen. Because models PAa, PAc, PAd, A, and Aa rely on pseudo-analytical or analytical solutions for only one limiting substrate, they do not include a separate mass balance for dissolved oxygen. Incorporating oxygen limitation in these models involves applying $S_{LF,O2}$ for all positions in the biofilm; this simplification works well when the dissolved oxygen concentration does not decrease much in the biofilm or when $S_{LF,O2} >> K_{O2}$. However, in cases where the substrate flux is controlled by oxygen rather than by the donor substrate, the models based on the pseudo-analytical solution are adapted by having oxygen take the role of the limiting substrate. The limiting substrate is determined by computing all fluxes assuming either substrate is solely rate limiting. For example, in the special case of a high influent N:COD ratio, the roles of ammonia nitrogen and oxygen are reversed, because NH_4^+-N removal rates are lower assuming oxygen is the limiting substrate. In this case, the rate-limiting substrate is O_2, and the flux of NH_4^+-N is computed from stoichiometry.

One final distinction is between AQUASIM-based solutions (models N1, N1a and N1b) and other BM3 1d models. AQUASIM explicitly recognizes that diffusion occurs in the biofilm liquid phase only by distinguishing between the water and the solid phase using a porosity value ε_l (set at 0.8). The porosity is used to reduce the surface area for diffusion to $\varepsilon_l A_F$ in Equation (4.31). The other models do not make this change. A means to make the AQUASIM solution more consistent with the other four models is to increase the diffusion coefficients in the biofilm (D_F) by $1/\varepsilon_l$, as in models N1a and N1b.

4.4.4 Results from one-dimensional models

4.4.4.1 Standard case

Table 4.20 summarizes the key output results for the standard case. It also shows the format for presenting all the results for BM3. Each row represents one model identified its letter code. The codes and order are the same as in Table 4.19, which summarizes the key characteristics of each model. The columns provide the bulk-liquid concentrations (g/m^3) of substrate (COD) and NH_4^+-N, the fluxes ($g/m^2 d$) for COD and NH_4^+-N, and the average surface coverage (g_{COD-X}/m^2) of heterotrophs, nitrifiers, and inert biomass. The last row gives the mean value for eight results in each column; the result for Aa is not included in the mean, because it was systematically different from the other. In the table, entries in **boldface** type are noticeably larger than the mean, while values in *italics* are smaller.

The first key difference illustrated in Table 4.20 is the distribution of the three types of biomass. Models N1, N1a, and PAc show substantially more nitrifiers and inerts than do the other models. To balance, the heterotrophs are less. Models N1, N1a, and PAc are the ones that *protect* the slow-growing species by having them migrate to the back of the biofilm. This migration occurs naturally due to the way surface detachment is implemented in models N1 and N1a, while layering is imposed *a priori* in model PAc.

When the nitrifiers are not protected by being near the substratum, their mass is much lower, which results in a significant decrease in the NH_4^+-N flux and increase in the bulk concentration of NH_4^+-H, as indicated for models N1b, N1s, and PAd. These three models impose an even, average biomass distribution for all locations in the biofilm. For the standard case, in which all biomass types are present, allowing for protection of slow growers increases the amount of nitrifier biomass, which results in more NH_4^+-N removal.

Table 4.20. Summary of key output parameters for the standard case. For each column, the **boldface** entries are larger than the mean, while *italics* entries are smaller than the mean. The column mean does not include the result from model Aa.

Model	$S_{B,S}$ g_{COD_S}/m^3	$S_{B,N}$ gN/m^3	$j_{F,S}$ g_{COD_S}/m^2d	$j_{F,N}$ g_N/m^2d	Heterotrophs g_{COD_X}/m^2	Nitrifiers g_{COD_X}/m^2	Inerts g_{COD_X}/m^2
N1	**5.39**	*1.59*	*4.92*	0.88	*1.88*	**0.79**	**2.33**
N1a	4.84	*1.45*	**5.03**	**0.91**	*2.02*	**0.83**	**2.15**
N1b	4.94	**2.30**	*5.01*	*0.74*	**3.91**	*0.32*	*0.77*
N1s	4.96	**2.32**	*5.01*	*0.74*	**3.91**	*0.31*	*0.78*
A	**5.20**	*1.76*	*4.96*	**0.85**	**3.86**	*0.36*	*0.77*
Aa	*2.11*	*0.94*	**5.58**	**1.00**	**3.96**	*0.38*	*0.66*
PAa	*4.36*	*1.62*	**5.13**	**0.88**	**3.90**	*0.36*	*0.74*
PAc	*4.39*	*1.65*	**5.12**	**0.87**	*2.23*	**0.84**	**1.93**
PAd	4.94	**2.20**	*5.01*	*0.76*	**3.90**	*0.33*	*0.77*
Mean	4.87	1.87	5.02	0.83	3.20	0.52	1.28

The higher COD fluxes (and lower S_S) with models PAa, PAc, and Aa are a second key difference. They occur for models PAa and PAc because these models avoid any *dilution* of the heterotrophs with nitrifiers or inerts near the outer surface of the biofilm. Model PAa does this by making the calculations of heterotroph and nitrifier flux independent of each other, while model PAc places all nitrifiers and inerts in layers behind the heterotrophs. Model Aa gives a higher flux of COD (and NH_4^+-N, too) because it uses zero-order kinetics throughout the biofilm. This trend of Aa having higher fluxes is common for all cases and is inherent to the zero-order solution. This trend is not pointed out again unless it has special interest.

Third, models N1b, N1s, and PAd predicted similar results for all parameters, and this reflects that each of them distributed the biomass types evenly across the biofilm. Although each of the models achieved an even biomass distribution by a different strategy, the model predictions were quite similar. These three models gave similar results for most of the special cases, and this similarity is pointed out again only when of special interest.

Finally, model N1 has lower COD and NH_4^+-N fluxes than does model N1a, even though both have similar distributions of biomass. The smaller fluxes with model N1 are due to the lowering of the effective surface area for transport when a biofilm porosity of 0.8 is input to AQUASIM. This flux trend appears for COD and NH_4^+-N in all the special cases and is not highlighted again. It is discussed in more detail for the first benchmark problem.

4.4.4.2 High influent N:COD

Table 4.21 summarizes the outputs for the first special case of a high influent N:COD ratio, achieved by increasing the influent NH_4^+-N concentration from 6 to 30 g_N/m^3. The high N:COD ratio accentuates any interaction that is controlled by the presence of nitrifiers.

The surface coverage of heterotrophs is smaller for the three models that protect slower growing species by having them accumulate near the substratum. Among the three models, model PAc favors the accumulation of nitrifiers more than does models N1 and N1a, because it places all of the nitrifiers in a protected layer behind the heterotrophs. Although model PAc has the highest accumulation of nitrifier biomass (and inerts), its NH_4^+-N flux is not as large as for models N1 and N1a. This reduction in $j_{F,N}$ occurs because of the *added mass-transport resistance* to get NH_4^+-N across the heterotroph layer to reach the nitrifiers.

Table 4.21. Summary of key output parameters for the special case of a high influent N:COD ratio

Model	$S_{B,S}$ g_{CODS}/m^3	$S_{B,N}$ gN/m^3	$j_{F,S}$ g_{CODS}/m^2d	$j_{F,N}$ g_N/m^2d	Heterotrophs g_{CODX}/m^2	Nitrifiers g_{CODX}/m^2	Inerts g_{CODX}/m^2
N1	**5.86**	18.93	*4.83*	2.21	*1.73*	1.07	**2.20**
N1a	**5.35**	17.03	4.93	2.59	*1.83*	1.24	**1.93**
N1b	5.12	**20.61**	4.98	*1.88*	**3.68**	*0.72*	*0.61*
N1s	5.13	**20.83**	4.97	*1.83*	**3.67**	*0.70*	*0.63*
A	**5.54**	17.59	4.91	2.48	**3.46**	0.93	*0.61*
Aa	*2.27*	17.76	**5.55**	2.45	**3.63**	*0.82*	*0.55*
PAa	*4.36*	*6.74*	**5.13**	**4.65**	3.12	1.42	*0.46*
PAc	*4.47*	**19.77**	**5.11**	2.05	*1.82*	**1.74**	**1.44**
PAd	5.21	18.41	4.96	2.32	**3.51**	0.87	*0.62*
Mean	5.13	17.49	4.98	2.50	2.86	1.09	1.07

See Table 4.20 for notes.

For the heterotrophs, models PAa, PAc, and Aa have the highest substrate fluxes and lowest bulk COD concentrations. The first two models have higher $j_{F,S}$ values because the heterotrophs in the outer layer are *not diluted* in any way by nitrifiers or inert biomass. Increasing the accumulation of nitrifiers in this special case accentuates the dilution effect on heterotrophs near the outer surface. Model Aa has a higher $j_{F,S}$ value because it used zero-order kinetics, and its $j_{F,N}$ value is higher than all other $j_{F,N}$ values for the same reason.

4.4.4.3 Low Influent N:COD

Table 4.22 summarizes the results for a low influent N:COD ratio, achieved by lowering the influent NH_4^+-N to 1.2 g_N/m^3. A low N:COD ratio reduces the importance of nitrifiers, making them much more susceptible to impacts from heterotrophs and inerts.

The most dramatic effect shown in Table 4.22 is that the nitrifiers are present only when protected by being near the substratum. The effect is stronger with models N1 and N1a, compared to model PAc. The difference is caused by the added mass-transport resistance to transport NH_4^+-N across the heterotroph layer to reach the nitrifier layer in model PAc, leading to a lower $j_{F,N}$ and a higher S_N. Another factor is the "degree of protection" imposed *a priori* in model PAc. All modeling cases make b_{de} for the inner, protected layers 10% of b_{de} for the outer heterotroph layer. A smaller ratio would protect the nitrifiers more and make the results of model PAc closer to those of models N1 and N1a. Model Aa also allows some nitrifier accumulation due to its faster zero-order kinetics.

Compared to models N1 and N1a, model PAc has a higher substrate (COD) flux and lower bulk concentration of COD, since the nitrifiers and inerts do not dilute the heterotrophs at all. This is why PAc matches PAa for substrate COD, even though nitrifiers are present with PAc, but "washed out" with PAa.

With nitrifiers "washed out of the biofilm," models N1b, N1s, and PAd give almost the same results for substrate COD. On the other hand, model PAa, which also has no nitrifiers, has a larger $j_{F,S}$ and a smaller S_S. This occurs because the inerts do not dilute the density of the heterotrophs, as they do for models N1b, N1s, and PAd. Model A, which has no nitrification, has a slightly smaller $j_{F,S}$, probably caused by the flux-averaging scheme.

Table 4.22. Summary of key output parameters for the special case of a low influent N:COD ratio

Model	$S_{B,S}$ g_{CODS}/m^3	$S_{B,N}$ gN/m^3	$j_{F,S}$ g_{CODS}/m^2d	$j_{F,N}$ g_N/m^2d	Heterotrophs g_{CODX}/m^2	Nitrifiers g_{CODX}/m^2	Inerts g_{CODX}/m^2
N1	**5.19**	*0.48*	4.96	**0.14**	*2.00*	**0.21**	**2.80**
N1a	4.66	*0.45*	5.07	**0.15**	*2.14*	**0.21**	**2.65**
N1b	4.81	**1.17**	5.04	*0.01*	**4.12**	*0.01*	*0.87*
N1s	4.82	**1.20**	5.04	*0.00*	**4.12**	*0.00*	*0.88*
A	**5.06**	**1.18**	4.99	*0.00*	**4.11**	*0.00*	*0.89*
Aa	*2.01*	*0.14*	**5.60**	**0.20**	**4.17**	0.08	*0.75*
PAa	*4.36*	**1.20**	5.13	*0.00*	**4.14**	*0.00*	*0.86*
PAc	*4.36*	*0.53*	5.13	**0.13**	*2.53*	**0.14**	**2.33**
PAd	4.81	**1.20**	5.04	*0.00*	**4.12**	*0.00*	*0.88*
Mean	4.76	0.93	5.05	0.05	3.41	0.07	1.52

See Table 4.20 for notes.

4.4.4.4 Low Production Rate for Inert Biomass

Table 4.23 summarizes the results for the special case in which the production of inert biomass is accentuated by increasing the fraction of decay that inactivated biomass to inerts from 20% to 50%. Respiration is decreased from 80% to 50% to compensate.

The most dramatic and obvious result is that models N1, N1a, and PAc show by far the largest increases in inert biomass, since they protect it near the substratum. These models also protect the nitrifiers and have the highest accumulations of them. The higher accumulations for nitrifiers give models N1, N1a, and PAc the greatest removals of NH_4^+-N. Similar to the special case with the high N:COD ratio (Table 4.21), models PAa and PAc have greater removals of substrate COD, because the large accumulation of inert biomass does not dilute the heterotrophs in the outer part of the biofilm. Whereas the buildup of nitrifiers dilutes the heterotrophs for the other models in the condition of high N:COD, the extra inerts dilute the heterotrophs in this special case.

Table 4.23. Summary of key output parameters for the special case of large production of inerts

Model	$S_{B,S}$ g_{CODS}/m^3	$S_{B,N}$ gN/m^3	$j_{F,S}$ g_{CODS}/m^2d	$j_{F,N}$ g_N/m^2d	Heterotrophs g_{CODX}/m^2	Nitrifiers g_{CODX}/m^2	Inerts g_{CODX}/m^2
N1	**5.71**	*1.72*	*4.86*	**0.86**	*1.74*	**0.69**	**2.57**
N1a	5.14	*1.56*	4.97	**0.89**	*1.87*	**0.72**	**2.41**
N1b	5.28	**3.06**	4.94	*0.59*	**3.42**	*0.21*	*1.36*
N1s	5.30	**3.07**	4.94	*0.59*	**3.42**	*0.21*	*1.37*
A	**5.51**	2.29	4.90	0.74	**3.38**	*0.27*	*1.35*
Aa	*2.35*	*1.13*	**5.53**	**0.97**	**3.49**	*0.31*	*1.20*
PAa	*4.36*	2.61	**5.13**	0.68	**3.45**	*0.24*	*1.31*
PAc	*4.53*	*1.49*	**5.09**	**0.90**	*1.68*	**0.73**	**2.59**
PAd	5.28	**2.92**	4.94	*0.62*	**3.42**	*0.22*	*1.36*
Mean	5.14	2.34	4.97	0.73	2.79	0.41	1.79

See Table 4.20 for notes.

4.4.4.5 High Detachment for a Thin Biofilm

Table 4.24 shows the outputs when the detachment rate is increased so that the biofilm is much thinner (20 μm), compared to all the other cases (500 μm). Two conclusions are obvious in Table 4.24. First, nitrifiers and inerts are totally washed out of the biofilm when the detachment rate is high enough to give a 20-μm biofilm. Second, all models except Aa have virtually the same results for removal of substrate COD when the dilution effect from nitrifiers or inerts is completely eliminated.

Table 4.24. Summary of key output parameters for the special case of high detachment leading to a thin biofilm (20 μm)

Model	$S_{B,S}$ g_{COD_S}/m^3	$S_{B,N}$ gN/m^3	$j_{F,S}$ g_{COD_S}/m^2d	$j_{F,N}$ g_N/m^2d	Heterotrophs g_{COD_X}/m^2	Nitrifiers g_{COD_X}/m^2	Inerts g_{COD_X}/m^2
N1	22.23	6.00	1.55	0.00	0.20	0.00	0.00
N1a	22.23	6.00	1.55	0.00	0.20	0.00	0.00
N1b	22.23	6.00	1.55	0.00	0.20	0.00	0.00
N1s	22.23	6.00	1.55	0.00	0.20	0.00	0.00
A	22.22	6.00	1.56	0.00	0.20	0.00	0.00
Aa	*20.80*	6.00	**1.84**	0.00	0.20	0.00	0.00
PAa	22.06	6.00	1.59	0.00	0.20	0.00	0.00
PAc	22.06	6.00	1.59	0.00	0.20	0.00	0.00
PAd	22.06	6.00	1.59	0.00	0.20	0.00	0.00
Mean	22.16	6.00	1.57	0.00	0.20	0.00	0.00

See Table 4.20 for notes.

4.4.4.6 Oxygen Sensitivity by Nitrifiers

Table 4.25 presents the results from the special case in which the nitrifiers' oxygen half-maximum-rate coefficient ($K_{O2,N}$) is increased from 0.5 to 5.0 g_{O2}/m^3. This condition seriously slows the nitrifiers' growth rate whenever the dissolved oxygen concentration is low in the biofilm. Comparing the results in Table 4.25 to those in Table 4.20 is very instructive.

Table 4.25. Summary of key output parameters for the special case of high oxygen sensitivity of nitrifiers

Model	$S_{B,S}$ g_{COD_S}/m^3	$S_{B,N}$ gN/m^3	$j_{F,S}$ g_{COD_S}/m^2d	$j_{F,N}$ g_N/m^2d	Heterotrophs g_{COD_X}/m^2	Nitrifiers g_{COD_X}/m^2	Inerts g_{COD_X}/m^2
N1	**5.41**	*1.87*	4.92	**0.83**	*1.84*	**1.21**	**1.95**
N1a	4.86	*1.68*	5.03	**0.86**	*1.98*	**1.21**	**1.81**
N1b	4.85	**4.92**	5.03	*0.22*	**4.05**	*0.11*	*0.84*
N1s	4.83	**5.81**	5.03	*0.04*	**4.11**	*0.02*	*0.88*
A	**5.12**	**4.07**	4.98	*0.39*	**4.00**	*0.17*	*0.83*
Aa	*2.10*	*1.31*	**5.58**	**0.94**	3.98	0.35	*0.67*
PAa	*4.36*	3.70	5.13	0.46	**4.00**	*0.21*	*0.79*
PAc	*4.40*	*1.88*	5.12	**0.83**	2.20	**0.91**	**1.89**
PAd	4.88	**4.08**	5.02	0.38	**4.00**	*0.18*	*0.82*
Mean	4.84	3.50	5.03	0.50	3.27	0.50	1.22

See Table 4.20 for notes.

Table 4.26. Summary of key output parameters for models PAa and PAc when $K_{O2,N}$ for NH_4^+-N = 5 g_{O2}/m^3, but the $K_{O2,N}$ value for endogenous respiration remains 0.5 $g_{O2}.m^3$

Model	$S_{B,S}$ g_{CODS}/m^3	$S_{B,N}$ gN/m^3	$j_{F,S}$ g_{CODS}/m^2d	$j_{F,N}$ g_N/m^2d	Heterotrophs g_{CODX}/m^2	Nitrifiers g_{CODX}/m^2	Inerts g_{CODX}/m^2
PAa	4.36	4.56	5.13	0.29	4.06	0.13	0.82
PAc	4.39	2.01	5.12	0.80	2.26	0.77	1.97

The negative impact of oxygen sensitivity is shown strongly by the outputs of models N1b, N1s, A, PAa, and PAd. With these models, the accumulation of nitrifiers declines noticeably from the standard condition (Table 4.20), and the removals of NH_4^+-N decline accordingly. On the other hand, the accumulation of nitrifiers actually increases for models N1, N1a, and PAc, which protect the nitrifiers by having them accumulate away from the outer surface of the biofilm. This result may seem counter-intuitive, because the dissolved oxygen concentration is lower inside the biofilm, which should make the effects of oxygen limitation even more severe for nitrifiers located deep inside the biofilm. Despite appearing to be counter-intuitive, the enhancement to nitrifier accumulation is a natural outcome of the way that the models are formulated. In particular, Table 4.16 shows that the respiratory decay rate is slowed by low S_{O2} in the same manner as are the NH_4^+-N oxidation and nitrifier synthesis rates. When the nitrifiers are deep inside the biofilm, their main loss mechanism is respiratory decay, not detachment. Therefore, increased sensitivity to low dissolved oxygen slows the main loss rate. The net effect is to increase the accumulation of nitrifiers when being near the substratum protects them. When not protected, detachment remains the nitrifiers' major loss mechanism. Because detachment is not affected by S_{O2} in models N1b, N1s, A, PAa, and PAd, more nitrifiers are lost from the biofilm when the synthesis rate is slowed by oxygen sensitivity.

A very different result occurs if the models have different $K_{O2,N}$ values for NH_4^+-N oxidation and endogenous respiration. This scenario seems more realistic, since the K_{O2} value for ammonia monooxygenase is likely to be much more oxygen sensitive than are the decay-related values (Malmstead *et al.* 1995; Rittmann and McCarty 2001). If only the $K_{O2,N}$ value for NH_4^+-N oxidation is increased, the nitrifiers are significantly disadvantaged by being away from the outer surface and, therefore, exposed to lower dissolved-oxygen concentrations. This is illustrated in Table 4.26, which shows the output of models PAa and PAc when only $K_{O2,N}$ for NH_4^+-N equals 5.0 g_{O2}/m^3, while the $K_{O2,N}$ value for endogenous respiration remains at 0.5 g_{O2}/m^3. In this case, the surface coverage of nitrifiers, as well as the removals of NH_4^+-N, declined in comparison to Tables 4.20 and 4.25. Thus, increased oxygen sensitivity for the oxidation of NH_4^+-N creates a competitive disadvantage for the nitrifiers for all models.

4.4.5 Lessons learned from the 1d BM3 models

BM3 was solved by nine different models. Four models (N1, N1a, N1b, and N1s) involve numerical solution of the differential mass balances for the various types of biomass and their substrates (COD, NH_4^+-N, and O_2). Three models (PAa, PAc, and PAd) exploit the pseudo-analytical solution for a steady-state biofilm so that the solution is with a spreadsheet. Models A and Aa use analytical solutions implemented through a spreadsheet. The most important distinction among the models include the way the biomass types are distributed in the biofilm. Models N1 and N1a allow the biomass types to distribute themselves according to the interactions among surface detachment, net growth, and biomass velocity. These models tend to have the slower growing species (nitrifiers and inerts) accumulate closer to the substratum, while the heterotrophs are nearer the outer surface. Model PAc imposes a layered distribution with the heterotrophs on the outside, the nitrifiers behind them, and the inerts next to the substratum. Models N1b, N1s, A, Aa, and PAd distributed the types evenly throughout the biofilm. Model PAa treats each biomass type as if it were independent of the others.

The models that protect the slow growing species by having them accumulate away from the outer surface (models N1, N1a, and PAc) always have the largest surface coverage by nitrifiers and inerts, but the heterotroph coverage declines to compensate. In terms of the nitrifiers and nitrification, the importance of this protection feature is accentuated when the nitrifiers are at risk of washout from a low influent N:COD ratio, a high detachment rate, or high oxygen sensitivity of the nitrifiers. In these cases, the nitrifiers are totally or almost totally washed out of the biofilm in models that do not afford protection by having the nitrifiers be away from the outer surface.

Coverage by heterotrophs and removal of substrate COD was most strongly affected by dilution from nitrifiers and inerts near the outer surface. Models that do not allow the nitrifiers and inerts to dilute the heterotrophs significantly in the outer layer (models PAa and PAc) predict more removal of COD than do the other models. This difference is accentuated when the slower-growing species are favored by a high influent N:COD ratio or a high production rate of inerts. The effect is negligible when the slow growers do not accumulate much in the biofilm due to a high detachment rate.

Model Aa, which employed zero-order kinetics in the biofilm, almost always has the highest substrate fluxes and the lowest substrate concentrations.

The results of multi-species BM3 illustrate that a wide range of 1d models is capable of representing the important interactions that can occur in biofilms in which distinctly different types of biomass can co-exist. The choice of the model to use depends on the user's needs and the modeling situation. One key choice is between models that demand a full numerical solution versus those that can be implemented with a spreadsheet. A second choice concerns the way in which the biomass is distributed. By far the simplest approach is to assume that the biomass types are independent of each other, as in model PAa. This approach may work well when protection of a slow-growing species (like the nitrifiers) or dilution of a fast-growing species (like the heterotrophs) is not a major issue. When protection of a slow-growing species is critical to an accurate representation, then a model that accumulates the slow growers away from the outer surface is essential (e.g., models N1, N1a, and PAc here). When the dilution of a fast-growing species by slower growers is key, then a model that distributes the different biomass types throughout the biofilm is essential (e.g., models N1, N1a, N1b,A, N1s, and PAd).

4.4.6 Two-dimensional models applied

Two 2d models also are applied to BM3. The N2a model is described in Section 0 and N2f model in Section 3.6.4. In 2d, the governing differential equations for soluble substrates within the biofilm are:

$$D_i \left(\frac{\partial^2 S_i}{\partial x^2} + \frac{\partial^2 S_i}{\partial y^2} \right) + r_i = 0 \qquad \text{for } i = S, N \text{ and } O_2 \qquad (4.35)$$

$$S_i(L_F, y) = S_i(L_X, y) = S_{B,i} \qquad (4.36)$$

$$\frac{\partial S_i(0, y)}{\partial x} = 0 \qquad (4.37)$$

$$S_i(x,0) = S_i(x,L_Y) \tag{4.38}$$

where r_i is the net reaction rate of substrate i, as a function of the concentrations organic substrate (S_S), ammonia-nitrogen (S_N), and oxygen (S_{O2}), and the concentrations of heterotrophic (X_H) and autotrophic (X_N) biomass. The boundary conditions describe the substrate concentrations at the biofilm-liquid interface (at $x=L_F$) as equal to the bulk concentrations (at $x=L_X$, with $L_X=L_F$ at steady state) (Equation (4.35)), the gradient of substrates at the impermeable substratum equal to zero (Equation (4.37)), and lateral periodicity at the other edges of the 2D domain (Equation (4.38)). Equation (4.36) defines a boundary condition without a concentration boundary layer, as specified in the definition of BM3.

Biomass growth is modeled as a function of the location within the biofilm and obeys the general mass balances. A significant difference between the 2d and 1d models is the simulation of biomass spreading within the biofilm. While the 1d models employ the concept of a velocity at which different types of biomass move away from the substratum, the 2d models use a discretized representation of biomass as "microbial particles." These particles grow in mass according to the mass balances shown above, but are limited to a maximum density. When this maximum density is reached, the microbial particles divide, and the "newborn" microbial particle is placed in a random location next to the "mother" particle. With this discrete rule, biofilm growth occurs when newborn particles shove other particles away. Biomass detachment is a consequence of this shoving effect. When a microbial particle is pushed across the imposed biofilm thickness limit, the particle is removed from the biofilm.

The reactor's bulk-liquid volume is modeled as completely mixed, according to the general mass balance represented by Equation (4.39), in which L_x and L_y are the dimensions of simulated biofilm in directions perpendicular and parallel to the substratum, respectively, and A_F and V_B represent the total area of biofilm and the reactor volume (which is considered for simplicity the bulk liquid volume), respectively. Both 2d models are solved in a dynamic manner, with an initial condition defined by randomly seeding a small number of heterotrophic and autotrophic "microbial particles" at the substratum, and allowing the system to reach steady state.

$$\frac{dS_{B,i}}{dt} = \frac{Q}{V_B}\left(S_{in,i} - S_{B,i}\right) + \frac{A_F}{V_B L_Y} \int_0^{L_X} \int_0^{L_Y} r_i(x,y)dydx \quad \text{for } i = S, N \text{ and } O_2 \tag{4.39}$$

The N2a model is a hybrid differential-discrete particle-based model (Picioreanu et al. 2004) derived from the cellular-automaton approach of Picioreanu et al. (1998a) and the individual-based approach by Kreft et al. (2001). Substrate gradients are simulated on a continuum space. Biomass is distributed in spherical microbial particles that push each other when growing. A microbial particle could contain biomass from of any type (i.e., heterotrophic, autotrophic, or inert), but is restricted to a maximum concentration of $C_{X,max} = 10000$ g/m^3, which corresponds to the maximum biomass density specified for BM3. In a particle-division event, the "newborn" particle randomly receives between 40 and 60% of the biomass contained in the "mother" particle.

The N2f model is a fully discretized cellular automaton model based on Pizarro et al. (2001) and Noguera et al. (2004). In this model, differential equations are not used to describe substrate concentration and transport nor biomass growth. Rather, stochastic discrete rules about local "food particle" and "microbial particle" movement and fate simulate mass transport, substrate utilization, and microbial growth and distribution within the biofilm. Substrate diffusion is implemented using the concept of random walks of food particles (Chopard and Droz 1991; Pizarro et al. 2001), while substrate utilization and microbial growth are represented by a discrete version of Monod kinetics (Noguera et al. 2004). Microbial particles are defined of being of three different types (i.e., heterotrophic, autotrophic, or inert) and having a concentration equal to the maximum biomass density within the biofilm. Decay events convert heterotrophic and autotrophic particles into inactive particles, while a division event creates a "newborn" particle with exactly the same characteristics as the "mother" particle.

4.4.7 Results for the two-dimensional models

The results of the 2d models are compared to each other and to the results from the 1d models that used fully numerical solutions (N1 and N1a models). Table 4.27 summarizes the key output parameters for the standard case. As a measurement of similarity in the output parameters, the last two rows in the table indicate average values and the relative variation for each parameter. The latter is calculated as the standard deviation divided by the average, and expressed as a percentage. For parameters with relative variations greater than 20%, entries significantly larger than the mean (greater than 1.2 times the average) are indicated in **boldface**, while those significantly lower than the mean (smaller than the average divided by 1.2) are indicated in *italics* font.

Table 4.27. Summary of key output parameters for the standard case with 2d models

Model	$S_{B,S}$ g_{CODS}/m^3	$S_{B,N}$ gN/m^3	$j_{F,S}$ g_{CODS}/m^2d	$j_{F,N}$ g_N/m^2d	**Heterotrophs** g_{CODX}/m^2	**Nitrifiers** g_{CODX}/m^2	**Inerts** g_{CODX}/m^2
N2a	5.14	1.50	4.95	0.89	1.81	0.72	**2.60**
N2f	5.14	1.74	4.96	0.85	**2.88**	0.68	*1.44*
N1	5.39	1.59	4.92	0.88	1.88	0.79	2.33
N1a	4.84	1.45	5.03	0.91	2.02	0.83	2.15
Average	5.13	1.57	4.97	0.88	2.15	0.76	2.13
Relative variation	4.4%	8.1%	0.9%	2.8%	23.1%	9.0%	23.3%

According to Table 4.27, the four models have excellent agreement in the prediction of bulk substrate concentrations and fluxes into the biofilm, which are two main parameters of interest in a macroscopic analysis of biofilm activity. The most significant deviations among the model output parameters are the higher accumulation of heterotrophic biomass in the N2f model and the higher accumulation of inert biomass in the N2a model. The N2f model also results in a significantly lower accumulation of inert biomass. However, these differences in biomass accumulation do not significantly influence the calculation of fluxes or bulk concentrations, likely due to the depletion of organic substrate in the deeper regions of the biofilm, which contributes to the very low activity of a large fraction of the heterotrophic biomass. Furthermore, the variation in predictions between the 2d and 1d models shown in Table 4.27 is not as large as the variations observed with the other 1D models (section 4.4.4.1).

The same trends of higher heterotrophic biomass accumulation in the N2f model and higher accumulation of inerts in the N2a model, but insignificant differences in the predictions of substrate concentrations and fluxes are observed in the simulations of the special case with a high N:COD ratio (Table 4.28). Notably, all the models agree in the accumulation of nitrifiers, which is the bacterial group enhanced in this special case.

Table 4.28. Summary of key output parameters for the special case of high influent N:COD ratio with 2d models

Model	$S_{B,S}$ g_{CODS}/m^3	$S_{B,N}$ gN/m^3	$j_{F,S}$ g_{CODS}/m^2d	$j_{F,N}$ g_N/m^2d	Heterotrophs g_{CODX}/m^2	Nitrifiers g_{CODX}/m^2	Inerts g_{CODX}/m^2
N2a	5.45	18.15	4.90	2.35	1.71	1.07	**2.42**
N2f	5.56	20.26	4.89	1.89	**2.92**	1.10	*0.98*
N1	5.86	18.93	4.83	2.21	1.73	1.07	2.20
N1a	5.35	17.03	4.93	2.59	1.83	1.24	1.93
Average	5.56	18.59	4.89	2.26	2.05	1.12	1.88
Relative variation	4.0%	7.3%	0.9%	12.9%	28.5%	7.3%	33.7%

The models also agree in the relative distribution of biomass within the biofilm, as depicted in Figure 4.12, which compares output profiles from the N2a and N1a models. For this special case, inert biomass accumulates in the deeper region of the biofilm, while the majority of heterotrophs are located in the outer layers of the biofilm. The biomass distribution predicted with the N2f was slightly different as shown in Figure 4.13. This model gives significantly higher survival of heterotrophs in the deeper layers of the biofilm, and the distribution of nitrifiers and inactive biomass is even throughout the biofilm. The differences in biomass distribution with this model are likely the due to the rules used for biomass spreading.

When the special case of low N:COD ratio is simulated, the N2f model predicts a significantly lower accumulation of nitrifiers compared to the other models (Table 4.29), suggesting that the dynamics of biofilm growth in this model do not offer the same level of "protection" to this slow-growing population as the other models do. Nevertheless, the trend of similar predictions of bulk substrate concentrations and fluxes is maintained in this special case. Furthermore, the prediction in nitrifier accumulation is not as extreme as observed with the other 1d models, for which the nitrifier population disappeared (section 4.4.4).

Table 4.29. Summary of key output parameters for the special case of low influent N:COD ratio with 2d models

Model	$S_{B,S}$ g_{CODS}/m^3	$S_{B,N}$ gN/m^3	$j_{F,S}$ g_{CODS}/m^2d	$j_{F,N}$ g_N/m^2d	Heterotrophs g_{CODX}/m^2	Nitrifiers g_{CODX}/m^2	Inerts g_{CODX}/m^2
N2a	4.39	0.44	5.16	0.15	2.11	0.23	2.73
N2f	4.98	0.48	4.96	0.12	2.96	*0.13*	1.91
N1	5.19	0.48	4.96	0.14	2.00	0.21	2.80
N1a	4.66	0.45	5.07	0.15	2.14	0.21	2.65
Average	4.81	0.46	5.04	0.14	2.30	0.20	2.52
Relative variation	7.3%	4.5%	1.9%	10.1%	19.2%	22.7%	16.4%

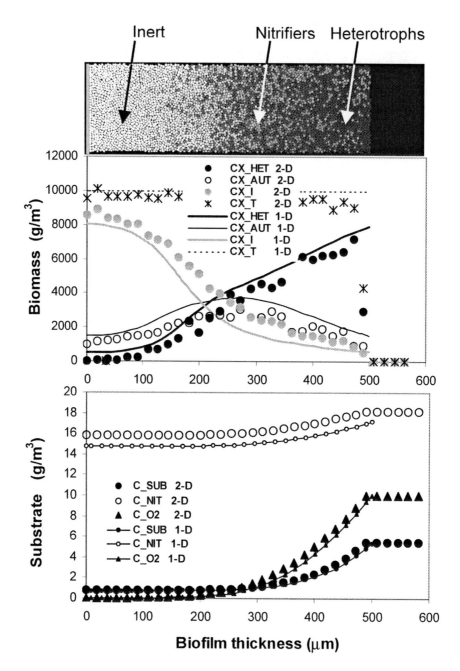

Figure 4.12. Biomass distributions in the biofilm at steady state (120 days) for the Special Case of High Influent N:COD. Lines in the graph are results of 1d N1a model, and symbols are average profiles computed with N2a particle-based 2d model. The top figure shows the 2d biomass distribution, with heterotrophic (dark gray) and nitrifying (light gray) particles. The whiter the color of the particles, the more inerts they contain (e.g., in biofilm depth – here to the left of the image).

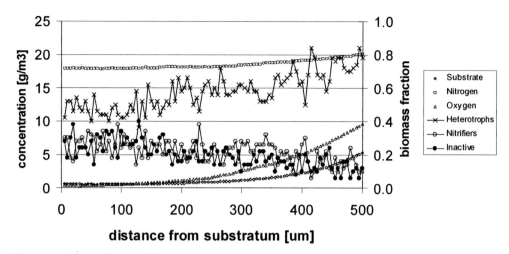

Figure 4.13. Results of biomass distribution and substrate profiles within the biofilm from the N2f model for the Special Case of High Influent N:COD.

When the production of inerts is accentuated by increasing the inactivation rate, the models exhibit the highest degree of variation in the output parameters (Table 4.30). This is the only simulation in which the predictions of substrate concentrations have a relative variation greater than 20%, with the N2a model providing a significantly higher nitrogen concentration, in agreement with its prediction of a significantly lower nitrifier population. In contrast, the N1a model predicts a significantly lower nitrogen concentration, but this prediction is also in agreement with having the highest accumulation in the nitrifier population. The N1 model results in the highest accumulation of inerts and a significantly lower accumulation of heterotrophic biomass.

Table 4.30. Summary of key output parameters for the special case of large production of inerts with 2d models

Model	$S_{B,S}$ g_{CODS}/m^3	$S_{B,N}$ gN/m^3	$j_{F,S}$ g_{CODS}/m^2d	$j_{F,N}$ g_N/m^2d	**Heterotrophs** g_{CODX}/m^2	**Nitrifiers** g_{CODX}/m^2	**Inerts** g_{CODX}/m^2
N2a	4.61	**3.03**	5.10	0.60	**2.80**	*0.24*	2.03
N2f	5.54	1.97	4.92	0.81	2.54	0.49	1.97
N1	5.71	1.72	4.86	0.86	*1.74*	**0.69**	2.57
N1a	5.14	*1.56*	4.97	0.89	1.87	**0.72**	2.41
Average	5.25	2.07	4.96	0.79	2.24	0.54	2.25
Relative variation	9.3%	32%	2.1%	16.6%	22.9%	41.4%	13.0%

The best agreement among the models is observed in the special case in which the detachment rate is accentuated, giving a thin biofilm, 20 μm (Table 4.31). The only significant difference is the accumulation of a small amount of inactive biomass in the N2f model, which is likely the result of a few "inert microbial particles" that remained embedded within the biofilm matrix.

Table 4.31. Summary of key output parameters for the special case of high detachment leading to a thin biofilm (20 μm) with 2d models

Model	$S_{B,S}$ g_{CODS}/m^3	$S_{B,N}$ gN/m^3	$j_{F,S}$ g_{CODS}/m^2d	$j_{F,N}$ gN/m^2d	**Heterotrophs** g_{CODX}/m^2	**Nitrifiers** g_{CODX}/m^2	**Inerts** g_{CODX}/m^2
N2a	22.19	6.00	1.55	0.00	0.20	0.00	0.00
N2f	21.89	6.00	1.34	0.00	0.20	0.00	0.003
N1	22.23	6.00	1.55	0.00	0.20	0.00	0.00
N1a	22.23	6.00	1.55	0.00	0.20	0.00	0.00
Average	22.14	6.00	1.50	0.00	0.20	0.00	0.00
Relative variation	0.7%	0.0%	7.0%		0.0%		200.0%

Finally, when the oxygen sensitivity of the nitrifiers is tested by increasing the oxygen half-saturation concentration (Table 4.32), the 2d models are consistently different from the 1d models in the prediction of biomass distribution, although the variations in predictions of bulk substrate concentrations and fluxes are again insignificant. In the 2d models, the lower rate of nitrogen utilization that results from an increase in the half-saturation concentration effectively reduced the amount of nitrifier biomass in the biofilm (compared to Table 4.27). The opposite result is observed for the 1d models, which predict an increase in the accumulation of nitrifiers. As discussed in section 4.4.4, the counter-intuitive increase in nitrifier population in the 1d models is due to a slowing of the endogenous decay rate. When the slow-growing nitrifiers are protected from detachment, they survive well when the decay rate declines in the same proportion as the synthesis rate. The 2d models, with their mechanisms of random placement of "newborn" microbial particles and the shoving of microbial particles in the internal regions of the biofilm did not result in the same kind of stable protection of the slow-growing nitrifiers.

Table 4.32. Summary of key output parameters for the special case of high oxygen sensitivity of nitrifiers with 2d models

Model	$S_{B,S}$ g_{CODS}/m^3	$S_{B,N}$ gN/m^3	$j_{F,S}$ g_{CODS}/m^2d	$j_{F,N}$ gN/m^2d	**Heterotrophs** g_{CODX}/m^2	**Nitrifiers** g_{CODX}/m^2	**Inerts** g_{CODX}/m^2
N2a	5.98	1.74	4.82	0.85	2.74	*0.52*	1.84
N2f	5.11	2.40	4.99	0.69	**2.88**	*0.65*	1.47
N1	5.41	1.87	4.92	0.83	*1.84*	**1.21**	1.95
N1a	4.86	1.68	5.03	0.86	1.98	**1.21**	1.81
Average	5.34	1.92	4.94	0.81	2.36	0.90	1.77
Relative variation	9.0%	17%	1.9%	9.8%	22.3%	40.6%	11.7%

4.4.8 Lessons learned from the 2d BM3 models

The general trend observed in the comparison of two 2d models with two 1d models is that all the models produce similar predictions for bulk substrate concentrations and fluxes into the biofilm, even though, in some cases, the models have significantly different accumulations of the different types of biomass. The similarity in output parameters for substrate concentration and fluxes is likely the result of having all the models being based on a similar theoretical and mathematical framework, with the obvious difference that substrate concentrations in the 2d models are calculated using mathematical descriptions that represent the 2d domain, while the 1d models have simpler 1d mathematical descriptions. Nevertheless, since the specifications of BM3 restrict the system to a flat biofilm in a completely mixed reactor, the problem is intrinsically a 1d problem not having prominent characteristics that can be captured only by a multidimensional model.

On the other hand, the predicted distributions of the different types of biomass varied considerably. In this regard, the only identifiable trend when comparing the 1d to the 2d models is the apparent increased, stable protection of nitrifiers in the 1d models, especially when this population is affected by a large accumulation of inerts (Table 4.30) or a lower rate of oxygen utilization (Table 4.32). This trend likely reflects the most significant difference between 1d and 2d models, which is the mechanism used to distribute the biomass within the biofilm. While the 1d models used the concept of a continuum field of biomass that moves away from the substratum, the 2d models uses the concept of microbial particles and cellular automaton rules to place newborn microbial particles close to the mother particles, with the final distribution of biomass within the biofilm being a consequence of the self-organization of the particles and the shoving of old particles by new particles.

Nomenclature

Dimensions are defined according to the SI base quantities L for length, M for mass, T for time, I for electric current, N for amount of substance, and θ for temperature. e^- stands for electronic charge.

Symbol	Definition	Dimension	Defined in Section or Equation
a	Specific surface area of the biofilm (biofilm surface/biofilm volume)	$L^2 L^{-3}$	Eq (2.20)
A_F	Biofilm surface area or area of the biofilm-liquid interface	L^2	Eq (2.25) § 4.2.1
A_S	Biofilm substratum area	L^2	Eq (2.32)
Ae	Biofilm area enlargement number	$L^2 L^{-2}$	§ 4.3.3.2
b	First-order biomass inactivation (decay) rate coefficient	T^{-1}	Eq (2.12)
b_H	First-order biomass decay rate constant for heterotrophic biomass	T^{-1}	§ 4.2.1
b_{de}	First-order biomass detachment rate constant	T^{-1}	§ 3.2.2
b_{ina}	First-order biomass inactivation rate constant	T^{-1}	§ 3.2.2
b_{res}	First-order biomass endogenous respiration rate constant	T^{-1}	§ 3.2.2
$B_{V,BOD}$	BOD$_5$ load per trickling filter volume	$M L^{-3} T$	Eq (1.1)
C_i	Concentration of a component i	$M_i L^{-3}$	Eq (2.16) Eq (3.3) § 2.4.3
$C_{B,i}$	Concentration of a component i in the bulk liquid	$M_i L^{-3}$	Eq (2.17)
$C_{in,i}$	Concentration of a component i in the influent	$M_i L^{-3}$	Eq (3.36)
$C_{ef,i}$	Concentration of a component i in the effluent	$M_i L^{-3}$	Eq (3.36)

© IWA Publishing 2006. *Mathematical Modeling of Biofilms: Scientific and Technical Report No.18* by the IWA Task Group on Biofilm Modeling (Hermann Eberl, Eberhard Morgenroth, Daniel Noguera, Cristian Picioreanu, Bruce Rittmann, Mark van Loosdrecht and Oskar Wanner). ISBN: 1843390876. Published by IWA Publishing, London, UK

$C_{LF,i}$	Concentration of a component i at the biofilm surface	$M_i\,L^{-3}$	Eq (2.17)
			Eq (3.22)
$C_{F,i}$	Concentration of a substrate i per unit biofilm liquid phase	$M_S\,L^{-3}$	Eq (3.87)
d_p	Diameter of a biofilm support particle	L	Eq (2.18)
D	Molecular diffusion coefficient of a component in water	$L^2\,T^{-1}$	Eq (2.16)
D_i	Molecular diffusion coefficient of a soluble component i	$L^2\,T^{-1}$	§ 3.2
D_T	Turbulent diffusion coefficient of a component	$L^2\,T^{-1}$	Eq (2.16)
D_F	Diffusion coefficient in the biofilm compartment	$L^2\,T^{-1}$	§ 2.5.1
$D_{F,i}$	Diffusion coefficient of a component i in the biofilm compartment	$L^2\,T^{-1}$	Eq (3.83)
D_L	Diffusion coefficient in pure water (or liquid)	$L^2\,T^{-1}$	§ 2.5.1
D_{O2}	Diffusion coefficient of oxygen	$L^2\,T^{-1}$	Eq (2.5)
D_S	Diffusion coefficient of organic substrate	$L^2\,T^{-1}$	Eq (2.5)
f	Correction factor for diffusion in biofilm relative to diffusion in pure water	-	§ 2.5.1
f_S°	Maximum fraction of electrons used for biomass synthesis reactions per electrons provided by the donor substrate	-	§ 2.5.4
F	Ratio of the flow rate approaching the trickling filter and the wastewater flow	-	Eq (1.1)
F	Faraday constant	$I\,T\,\bar{e}^{-1}$	Eq (2.16)
F_{ef}	Mass flow rate of component exiting the system with the effluent	$M\,T^{-1}$	Eq (3.29)
F_{in}	Mass flow rate of component entering the system with the influent	$M\,T^{-1}$	Eq (3.29)
F_{gen}	Mass flow rate of component generated in the system	$M\,T^{-1}$	Eq (3.29)
F_B	Mass flow rate of component transformed in bulk liquid	$M\,T^{-1}$	Eq (3.30)
F_F	Mass flow rate of component transformed in biofilm	$M\,T^{-1}$	Eq (3.30)
$F_{rem,S}$	Substrate mass removed per unit time in the whole system	$M_S\,T^{-1}$	Eq (2.29)
$F_{B,S}$	Overall substrate consumption rate in the bulk liquid	$M_S\,T^{-1}$	Eq (2.28)
$F_{F,S}$	Overall substrate consumption rate in the biofilm	$M_S\,T^{-1}$	Eq (2.28)
F_X	Total biomass accumulated in the system	$M_X\,T^{-1}$	Eq (2.34)
$F_{ef,X}$	Total flow rate of biomass exiting the system with the effluent	$M_X\,T^{-1}$	Eq (2.34)
$F_{in,X}$	Total flow rate of biomass entering the system with the influent	$M_X\,T^{-1}$	Eq (2.34)
$F_{B,X}$	Biomass suspended in the bulk liquid	$M_X\,T^{-1}$	Eq (2.34)
$F_{B,gro,X}$	Overall net biomass production rate (growth) in suspension	$M_X\,T^{-1}$	Eq (2.34)
$F_{F,X}$	Biomass accumulated in the biofilm	$M_X\,T^{-1}$	Eq (2.34)
$F_{F,gro,X}$	Overall net biomass production rate (growth) in biofilm	$M_X\,T^{-1}$	Eq (2.34)
F_{de}	Detachment speed function	$L\,T^{-1}$	Eq (3.117)
G	Dimensionless factor relating maximum bacterial growth rate in the biofilm to the maximum substrate transfer rate	-	Eq (2.26)
g	Liquid acceleration due to a body force	$L\,T^{-2}$	Eq (3.8)
I	Concentration of an inhibitor compound	$M_I\,L^{-3}$	Eq (2.10)
j	Vector of mass flux of a component	$M\,L^{-2}\,T^{-1}$	§ 2.4.3
			Eq (3.3)
j$_D$	Vector of diffusive flux of a component	$M\,L^{-2}\,T^{-1}$	Eq (3.4)
j$_C$	Vector of convective (advective) flux of a component	$M\,L^{-2}\,T^{-1}$	Eq (3.4)
j_n	Mass flux normal (perpendicular) to the biofilm surface	$M\,L^{-2}\,T^{-1}$	Eq (2.17)
j_F	Flux of component across the biofilm surface	$M\,L^{-2}\,T^{-1}$	Eq (3.32)
$j_{F,S}$	Overall flux of substrate at the biofilm surface (or net substrate flux consumed in the biofilm)	$M_S\,L^{-2}\,T^{-1}$	Eq (2.33)
			§ 4.2
$j_{F,O2}$	Overall flux of oxygen at the biofilm surface (or net oxygen flux consumed in the biofilm)	$M_{O2}\,L^{-2}\,T^{-1}$	§ 4.2
$j_{F,i}$	Flux of substrate i across the biofilm surface (with exponents (0) and (1) for zero and first order kinetics, respectively)	$M_{Si}\,L^{-2}\,T^{-1}$	§ 3.2.2
			Eq (3.83)
			Eq (3.96)

Symbol	Description	Units	Reference
$j_{M,i}$	Advective mass flux of particulate matter in the biofilm matrix	$M_{Xi}\, L^{-2}\, T^{-1}$	Eq (3.88)
$j_{F,S,i}$	Flux of a soluble component in a biofilm section i in the PN models	$M_S\, L^{-2}\, T^{-1}$	Eq (3.98)
$j_{F,S,tot}$	Overall specific flux of a soluble component in the PN models	$M_S\, L^{-2}\, T^{-1}$	Eq (3.98)
j_x, j_y, j_z	Mass flux of a component in the directions x, y or z	$M\, L^{-2}\, T^{-1}$	Eq (2.16) Eq (3.3)
k_H	Decay rate coefficient	T^{-1}	§ 4.2.1
k_0	Zero order reaction rate coefficient	$L^3\, M_S^{-1}\, T^{-1}$	Eq (2.8)
k_1	First order reaction rate coefficient	T^{-1}	Eq (2.7)
k_{at}	Attachment rate coefficient	T^{-1}	Eq (2.20)
$k_{at,i}$	Attachment mass transfer coefficient	LT^{-1}	Eq (3.94)
k_b	Backward reaction rate constant		§ 2.4.1
k_c	Liquid-biofilm mass transfer coefficient of a component	$L\, T^{-1}$	Eq (2.17)
k_{de}	Detachment rate coefficient	T^{-1}	Eq (2.20)
k_f	Forward reaction rate constant		§ 2.4.1
K_{eq}	Equilibrium constant		§ 2.4.1
K_I	Concentration of an inhibitor giving 50% inhibition of the rate	$M_I\, L^{-3}$	Eq (2.10)
K_{HCO3}	Monod half saturation constant for bicarbonate	$M_{HCO3}\, L^{-3}$	Eq (2.22)
K_{NH4}	Monod half saturation constant for ammonium	$M_{NH4}\, L^{-3}$	Eq (2.22)
K_{O2}	Monod half saturation constant for oxygen	$M_{O2}\, L^{-3}$	Eq (2.1) Eq (2.22)
K_S	Monod half saturation constant for substrate	$M_S\, L^{-3}$	Eq (2.1) Eq (2.6)
K_i	Monod half saturation constant for a substrate i	$M_i\, L^{-3}$	§ 3.2.2
K_{lim}	Monod half saturation constant for a limiting substrate	$M_{lim}\, L^{-3}$	§ 3.2.3.3
$K_{non\text{-}lim}$	Monod half saturation constant for a non-limiting substrate	$M_{non\text{-}lim}\, L^{-3}$	§ 3.2.3.4
L_F	Biofilm thickness	L	Eq (2.20) Eq (2.26)
L_L	Mass transfer boundary layer thickness	L	§ 2.4.3
$L_{F,i}$	Thickness of a biofilm section in the NP3 models	L	Eq (3.99)
$L_{F,avg}$	Average biofilm thickness	L	Eq (3.99) § 4.2
$L_{F,min}$	Minimum biofilm thickness (closest distance from biofilm substratum to the biofilm surface)	L	§ 4.2
$L_{F,max}$	Maximum biofilm thickness (largest distance from biofilm substratum to the biofilm surface)	L	§ 4.2
L_X, L_Y, L_Z	Dimensions of computational domain in multi-d models	L	§ 3.6.3.2.1
L_R, H_R	Reactor length and height in BM1, respectively	L	§ 4.2.1
$m_{X,i}$	Biomass component i in a biomass particle in models N2a	M_X	Eq (3.113)
$m_{X,max}$	Total maximum biomass in a particle at division, in models N2a	M_X	Eq (3.113)
m_X	Biomass contained in a biomass particle in CA models	M_X	Eq (3.120)
m_S	Substrate contained in a substrate particle in CA models	M_S	Eq (3.119)
M_X	Biomass in the reactor	M_X	§ 4.2.1
n_X	Number of particulate components	-	Eq (3.82)
n_S	Number of soluble components	-	
n_R	Number of reactions (processes) in which a component is involved	-	
n_P	Number of biomass particles in the N3e models	-	
p	pressure	$M\, L^{-1}\, T^{-2}$	Eq (3.8)
p_0, p_1, p_2	movement probabilities for particle in CA models	-	
p_r, p_g	reaction probabilities for particles in CA models	-	
p_{ina}, p_{res}			
$q_{max,i}$	Maximum specific conversion rate for a substrate i	$M_i\, M_X\, T^{-1}$	§ 3.2.2

$q_{max,i,mod}$	Modified maximum specific conversion rate for a substrate i	$M_{lim} M_X T^{-1}$	§ 3.2.3.3
Q	Volumetric flow rate	$L^3 T^{-1}$	Eq (2.25)
Q_{ef}	Effluent volumetric flow rate	$L^3 T^{-1}$	Eq (2.28)
Q_{in}	Influent volumetric flow rate	$L^3 T^{-1}$	Eq (2.28)
R	Universal gas constant	$L^2 M T^{-2} \theta^{-1} N^{-1}$	Eq (2.16)
R_p	Radius of a biomass particle in models N2a	L	Eq (3.113)
Re	Reynolds number (Re=UL/ν, with U a characteristic velocity and L a characteristic length)	-	Eq (2.19)
r_A	Net interfacial transfer rate of particulate components	$M_X L^{-2} T^{-1}$	Eq (2.20)
r_i	Overall (or net) transformation rate for component i	$M_i L^{-3} T^{-1}$	Eq (2.14) Eq (3.3)
r_F	Local rate of component transformation in the biofilm	$M L^{-3} T^{-1}$	Eq (3.33)
$r_{F,i}$	Net production rate of a soluble component in the biofilm	$M_S L^{-3} T^{-1}$	Eq (3.84)
$r_{M,i}$	Net production rate of a particulate component in the biofilm matrix	$M_X L^{-3} T^{-1}$	Eq (3.89)
$r_{B,i}$	Net production rate of a component i in the bulk liquid	$M_i L^{-3} T^{-1}$	Eq (3.95)
r_g	Rate of heterotrophic growth	$M_X L^{-3} T^{-1}$	Eq (2.1)
r_{in}	Rate of biomass inactivation or decay	$M_X L^{-3} T^{-1}$	Eq (2.12)
r_{O2}	Rate of oxygen utilization	$M_{O2} L^{-3} T^{-1}$	Eq (2.3)
r_S	Rate of substrate utilization	$M_S L^{-3} T^{-1}$	Eq (2.2), (2.13)
$r_{A,rem,S}$	Rate of substrate removal per unit biofilm area	$M_S L^{-2} T^{-1}$	Eq (2.32)
$r_{rem,S}$	Rate of substrate removal per unit system volume	$M_S L^{-3} T^{-1}$	Eq (2.31)
r_X	Rate of biomass production per unit volume	$M_X L^{-3} T^{-1}$	Eq (2.11)
$r_{A,pro,X}$	Overall rate of biomass production per unit biofilm substratum area	$M_X L^{-2} T^{-1}$	Eq (2.38) Eq (4.4)
$r_{pro,X}$	Overall rate of biomass production per unit system (reactor) volume	$M_X L^{-3} T^{-1}$	Eq (2.37)
Sc	Schmidt number (Sc=ν/D)	-	Eq (2.19)
Sh	Sherwood number (Sh=$k_c L/D$, with L a characteristic length)	-	Eq (2.18)
S	Concentration of a soluble component in general (e.g., substrate)	$M_S L^{-3}$	Eq (2.6)
S_i	Concentration of a soluble component i	$M_i L^{-3}$	§ 3.2.2
S_{ef}	Effluent substrate concentration	$M_S L^{-3}$	Eq (2.28)
S_{in}	Influent substrate concentration	$M_S L^{-3}$	Eq (2.28)
S_B	Substrate concentration in the bulk liquid	$M_S L^{-3}$	Eq (2.26)
$S_{in,S}$	Concentration of organic substrate (COD) in influent	$M_S L^{-3}$	§ 4.2.1
S_{O2}	Concentration of oxygen	$M_{O2} L^{-3}$	Eq (2.1) § 4.2.1
S_S	Concentration of organic substrate (COD)	$M_S L^{-3}$	Eq (2.1)
$S_{LF,O2}$	Concentration of oxygen at the biofilm surface	$M_{O2} L^{-3}$	Eq (2.5)
$S_{LF,S}$	Concentration of organic substrate (COD) at the biofilm surface	$M_S L^{-3}$	Eq (2.5)
$S_{LF,lim}$	Concentration of a limiting substrate i at the biofilm surface	$M_{lim} L^{-3}$	§ 3.2.3.3
$S_{LF,non-lim}$	Concentration of a non-limiting substrate i at the biofilm surface	$M_{non-lim} L^{-3}$	§ 3.2.3.4
$S_{LF,i}$	Concentration of a soluble substrate i at the biofilm surface	$M_i L^{-3}$	§ 3.2.2
$S_{LF,avg}$	Average concentration of substrate at the biofilm surface	$M_S L^{-3}$	Eq (4.5)
$S_{L,i}$	Concentration of a soluble component i in the liquid boundary layer	$M_i L^{-3}$	
$S_{B,S}$	Concentration of organic substrate (COD) in the bulk liquid	$M_S L^{-3}$	§ 4.2
$S_{B,O2}$	Concentration of dissolved oxygen in the bulk liquid	$M_{O2} L^{-3}$	§ 4.2
$S_{B,i}$	Concentration of a soluble substrate i in the bulk liquid	$M_i L^{-3}$	§ 3.2.2
$S_{F,i}$	Concentration of a soluble component i in the biofilm	$M_i L^{-3}$	Eq (3.83)
$S_{0,S}$	Concentration of organic substrate (COD) at the biofilm base (at the substratum)	$M_S L^{-3}$	§ 4.2

$S_{0,O2}$	Concentration of dissolved oxygen at the biofilm base (at the substratum)	$M_{O2}\,L^{-3}$	§ 4.2
t	Current time	T	Eq (3.3)
T	Temperature	θ	Eq (2.16) Eq (2.22)
U	Characteristic velocity in Re number	$L\,T^{-1}$	§ 2.4.3
u	Vector of advective velocity of a component or of liquid	$L\,T^{-1}$	Eq (2.16) Eq (3.7)
u_x, u_y, u_z	Advective velocity of a component in the directions x, y or z	$L\,T^{-1}$	Eq (2.16) Eq (3.6)
$u_{x\cdot max}$	Maximum liquid velocity at the top of the computational domain (BM2)	$L\,T^{-1}$	
u_F	Advective velocity of particulate components in the biofilm	$L\,T^{-1}$	Eq (3.88)
u_{at}	Global attachment velocity	$L\,T^{-1}$	Eq (3.92)
u_{de}	Global detachment velocity	$L\,T^{-1}$	Eq (3.92)
V	Biofilm reactor volume	L^3	Eq (2.31)
V_B	Bulk liquid volume	L^3	Eq (3.30)
V_F	Biofilm volume	L^3	Eq (3.33)
w_i	Weight given to a biofilm section i in the averaging in the NP3 models	-	Eq (3.98)
x	Vector of position	L	Eq (3.117)
x, y, z	Spatial directions. Usually z denotes the direction perpendicular to the substratum	L	Eq (2.16)
x_S	Degree of substrate conversion	-	Eq (2.30)
X	Concentration of the particulate material (microbial species or biomass type)	$M_X\,L^{-3}$	Eq (2.11)
X_B	Concentration of particulate component (e.g., biomass) in the bulk liquid	$M_X\,L^{-3}$	Eq (2.20)
X_{ef}	Concentration of particulate component in the system effluent	$M_X\,L^{-3}$	Eq (2.25) Eq (2.35)
X_{in}	Concentration of particulate component in the system influent	$M_X\,L^{-3}$	Eq (2.25) Eq (2.35)
X_F	Concentration of particulate component (e.g., biomass) in the biofilm (biomass per unit biofilm volume)	$M_X\,L^{-3}$	Eq (2.25)
$X_{F,A}$	Concentration of particulate component (e.g., biomass) in the biofilm per unit substratum area	$M_X\,L^{-2}$	Eq (2.35)
$X_{F,N}$	Concentration of autotrophic biomass in the biofilm	$M_X\,L^{-3}$	§ 3.2.3.2
$X_{F,H}$	Concentration of heterotrophic biomass in the biofilm	$M_X\,L^{-3}$	§ 3.2.3.2
$X_{F,I}$	Concentration of inert biomass in the biofilm	$M_X\,L^{-3}$	§ 3.2.3.2
$X_{F,max}$	Maximum possible concentration of biomass in the CA models	$M_X\,L^{-3}$	Eq (3.120)
X_H	Concentration of heterotrophic biomass (particulate component)	$M_X\,L^{-3}$	Eq (2.1)
$X_{I,i}$	Concentration at the bulk liquid side of the biofilm surface of the suspended particulate component i	$M_X\,L^{-3}$	Eq (3.94)
$X_{M,i}$	Concentration of a particulate component i (per volume biofilm)	$M_X\,L^{-3}$	Eq (3.80)
$X_{P,i}$	Concentration of suspended particulate components in the biofilm pores	$M_X\,L^{-3}$	Eq (3.82)
Y	Yield coefficient of biomass produced per unit substrate consumed	$M_X\,M_S^{-1}$	Eq (2.13)
Y_H	Yield of biomass produced on substrate utilized	$M_X\,M_S^{-1}$	Eq (2.2) § 4.2.1
Y_i	Yield of biomass produced on substrate i utilized	$M_X\,M_i^{-1}$	§ 3.2.2
$\gamma_{S,O2}$	Penetration of organic substrate relative to the penetration of oxygen in the biofilm	-	Eq (2.5)

δ	Substrate penetration depth in the biofilm	L	§ 3.2.2
ε_l	Ratio of liquid phase volume in the biofilm to total biofilm volume	$L^3 L^{-3}$	§ 2.5.1
$\varepsilon_{s,i}$	Volume fraction of a particulate component i in the biofilm (volume cells/volume biofilm)	$L^3 L^{-3}$	Eq (3.80)
$\varepsilon_{l,F}$	Volume fraction of liquid phase in the biofilm	$L^3 L^{-3}$	Eq (3.80)
ζ	Ion charge	$\bar{e} N^{-1}$	Eq (2.16)
θ	Biofilm porosity or pore volume fraction	$L^3 L^{-3}$	Eq (3.81)
μ	Specific biomass growth rate	T^{-1}	Eq (2.6)
μ_A	Specific growth rate of the autotrophic organisms	T^{-1}	Eq (2.22)
$\mu_{max,A,10}$	Maximum specific growth rate of autotrophic organisms at 10 °C	T^{-1}	Eq (2.22)
$\mu_{max,H}$	Maximum specific rate for heterotrophic growth	T^{-1}	Eq (2.1) § 4.2.1
μ_{max}	Maximum specific biomass growth rate	T^{-1}	Eq (2.6)
ν	Kinematic viscosity of the liquid	$L T^{-2}$	Eq (2.19) Eq (3.8)
$\nu_{S,O2}$	Stoichiometric coefficient substrate per oxygen utilized	$M_S M_{O2}^{-1}$	Eq (2.4)
ν_{ij}	Stoichiometric coefficients (multipliers) of component i in rate j	$M_i M_X^{-1}$	Eq (2.14)
ρ_j	Rate of process j	$M L^{-3} T^{-1}$	Eq (2.14)
ρ	Liquid density	$M L^{-3}$	Eq (3.8)
$\rho_{s,i}$	Density of a particulate component i (mass per cell volume)	$M_X L^{-3}$	Eq (3.80)
$\rho_{X,i}$	Density of a biomass component i in biomass particles in models N2a (per volume particle)	$M_X L^{-3}$	Eq (3.113)
$\sigma_X,$ $\sigma_Y, \sigma_Z,$ τ_{XY}	Stress components in a detachment model	$M L^{-1} T^{-2}$	Eq (3.115)
σ_e	Biofilm cohesion strength	$M L^{-1} T^{-2}$	Eq (3.115)
σ_t	Ultimate tensile strength	$M L^{-1} T^{-2}$	Eq (3.115)
τ_{dif}	Characteristic time for substrate diffusion	T	Eq (2.27)
τ_{gr}	Characteristic time for biomass growth	T	Eq (2.27)
τ_{LF}	Shear stress at the biofilm surface	$M L^{-1} T^{-2}$	Eq (3.93)
φ	Overall bed porosity (void volume fraction)	-	§ 2.4.3
Φ	Electrical potential	$M L^2 T^{-3} I^{-1}$	Eq (2.16)
$\tilde{\Phi}$	Dimensionless electrical potential	-	Eq (3.105)

References

Alpkvist, E. (2005) *Modelling and simulation of heterogeneous biofilm growth using a continuum approach.* Licensiate thesis in mathematical sciences, Lund University, Sweden (ISBN 91-631-6871-5).

Amann, R., Stromley, R., Devereaux, R., Key, R. and Stahl, D. (1992) Molecular and microscopic identification of sulfate-reducing bacteria in multispecies biofilms. *Appl. Environ. Microbiol.* **58**, 614-623.

Andrews, G. (1988) Effectiveness factors for bioparticles with Monod kinetics. *Chem. Eng. J.* **37**(2), B831-B837.

Atkinson, B. and Davies, I.J. (1974) The overall rate of substrate uptake (reaction) by microbial films. Part I. A biological rate equation. *Trans. Inst. Chem. Engrs.* **52**, 248-259

Atkinson, B. and How, S.Y. (1974) The overall rate of substrate uptake (reaction) by microbial films. Part II. Effect of concentration and thickness with mixed microbial films. *Trans. Inst. Chem. Engrs.* **52**, 260-272.

Bae, W. and Rittmann, B.E. (1996) A structured model of dual-limitation kinetics. *Biotechnol. Bioeng.* **49**, 683-689.

Batstone, D.J., Keller, J., Angelidaki, I., Kalyuzhnyi, S.V., Pavlostathis, S.G., Rozzi, A., Sanders W.T.M., Siegrist, H. and Vavilin, V.A. (2002) Anaerobic Digestion Model No 1 (ADM1), IWA Task Group for Mathematical Modelling of Anaerobic Digestion Processes, IWA Publishing, London, UK.

Beccari, M., Dipinto, A.C., Ramadori, R. and Tomei, M.C. (1992) Effects of dissolved-oxygen and diffusion resistances on nitrification kinetics. *Water Res.* **26**(8), 1099-1104.

Benham, P.P., Crawford, R.J. and Armstrong, C.G. (1996) *Mechanics of engineering materials.* Longman Grp. Ltd., Harlow, Essex, UK.

Beuling, E.E., Van den Heuvel, J.C. and Ottengraf, S.P.P. (2000) Diffusion coefficients of metabolites in active biofilms. *Biotechnol. Bioeng.* **67**(1), 53-60.

Bohlool, B. and Schmidt, E. (1980) The immunofluorescence approach in microbial ecology. *Adv. Microb. Ecol.* **4**, 203-236.

Bryers, J.D. and Drummond, F. (1998) Local macromolecule diffusion coefficients in structurally non-uniform bacterial biofilms using fluorescence recovery after photobleaching (FRAP). *Biotechnol. Bioeng.* **60**(4), 462-473.

Chang, I., Gilbert, E.S., Eliashberg, N. and Keasling, J.D. (2003) A three-dimensional, stochastic simulation of biofilm growth and transport-related factors that affect structure. *Microbiology SGM* **149**(10), 2859-2871.

Chen, S. and Doolen, G.D. (1998) Lattice Boltzmann method for fluid flows. *Annu. Rev. Fluid Mech.* **30**, 329-364.

Chopard, B. and Droz, M. (1991) Cellular automata model for the diffusion equation. *J Stat. Phys.* **64**(3/4), 859–892.

Chorin, A.J. (1967) A Numerical Method for Solving Incompressible Viscous Flow Problems. *Journal of Computational Physics*, **2**(1), 22-30.

Comiti, J., Mauret, E. and Renaud, M. (2000) Mass transfer in fixed beds: proposition of a generalized correlation based on an energetic criterion. *Chem. Eng. Sci.* **55**, 5545–5554.

Costerton, J.W., Geesey, G.G. and Cheng, K.J. (1978) How bacteria stick. *Sci. American* **238**, 86-95.

Costerton, J.W., Lewandowski, Z., Caldwell, D.E., Korber, D.R. and Lappin-Scott, H.M. (1995) Microbial biofilms. *Annu. Rev. Microbiol.* **49**, 711-745.

CRC (1992) *Handbook of Chemistry and Physics* (ed. D.R. Liede), CRC Press, Boca Raton, FL, USA.

De Beer, D., Stoodley, P. and Lewandowski, Z. (1997) Measurement of local diffusion coefficients in biofilms by microinjection and confocal microscopy. *Biotechnol. Bioeng.* **53**(2), 151-158.

Dillon, R., Fauci, L. and Gaver, D. (1995) A microscale model of bacterial swimming, chemotaxis and substrate transport. *J. Theor. Biol.* **177**, 325-340.

Dillon, R., Fauci, L., Fogelson, A. and Gaver, D. (1996) Modeling biofilm processes using the immersed boundary method. *J. Comput. Phys.* **129**, 57-73.

Dillon, R. and Fauci, L. (2000) A microscale model of bacterial and biofilm dynamics in porous media. *Biotechnol. Bioeng.* **68**, 536-547.

Dockery, J. and Klapper, I. (2001) Finger formation in biofilm layers. *SIAM J Appl. Math.* **62**(3), 853-869.

Dupin, H.J., Kitanidis, P.K. and McCarty, P.L. (2001) Pore-scale modeling of biological clogging due to aggregate expansion: A material mechanics approach. *Water Resources Research* **37**(12), 2965-2979.

D'Souza, R.M. and Margolus, N.H. (1999) A thermodynamically reversible generalization of Diffusion Limited Aggregation. *Phys. Review E*, **60**, 264-274.

Eberl, H.J., Picioreanu, C., Heijnen, J.J. and Van Loosdrecht, M.C.M. (2000a) A three-dimensional numerical study on the correlation of spatial structure, hydrodynamic conditions, and mass transfer and conversion in biofilms. *Chem. Eng. Sci.* **55**, 6209-6222.

Eberl, H.J., Picioreanu, C. and Van Loosdrecht, M.C.M. (2000b) Modeling geometrical heterogeneity in biofilms. In *High Performance Computing Systems & Applications* (eds. A. Pollard, D.J.K. Mewhort, D.F. Weaver), chapter 50, 497-512, Kluwer Academic Publishers.

Eberl, H.J., Parker, D.F. and van Loosdrecht, M.C.M. (2001) A new deterministic spatio-temporal continuum model for biofilm development. *J. Theor. Med.* **3**, 161-175.

Eberl, H.J., Efendiev, M.A. (2003) A Transient Density Dependent Diffusion-Reaction Model for the Limitation of Antibiotic Penetration in Biofilms. *Electronic Journal of Differential Equations* CS10, 123-142.

Efendiev, M.A., Eberl, H.J. and Zelik, S.V (2002) Existence and long time behaviour of solutions of a nonlinear reaction-diffusion system arising in the modeling of biofilms. *Nonlinear Diffusive Systems and Related Topics*, Res. Inst. Math. Sci. (Kyoto) **1258**, 49-71.

Fan, L.-S., Leyva-Ramos, R., Wisecarver, K.D. and Zehner, B.J. (1990). Diffusion of phenol through a biofilm grown on activated carbon particles in a draft-tube three-phase fluidized-bed bioreactor. *Biotechnol. Bioeng.* **35**, 279-286.

Finlayson, B.A. (1972) *The method of weighted residuals and variational principle; with application in fluid mechanics, heat and mass transfer*, Academic Press, New York, USA.

Flora, J.R.V., Suidan, M.T., Biswas, P. and Sayles, G.D. (1993) Modeling substrate transport into biofilms: Role of multiple ions and pH effects. *J. Environ. Eng.* **119**(5), 908-930.

Fogler, H.S. (1999) *Elements of chemical reaction engineering*. Upper Saddle River, Prentice-Hall PTR.

Frössling, N. (1938) *Gerlands Beitr.Geophys.* **52**, 170.

Fu, Y.-C., Zhang, T.C. and Bishop, P.L. (1994) Determination of effective oxygen diffusivity in biofilms grown in a completely mixed biodrum reactor. *Water Sci. Technol.* **29**(10/11), 455-462.

Furumai, H. and Rittmann, B.E. (1992) Advanced modeling of mixed populations of heterotrophs and nitrifiers considering the formation and exchange of soluble microbial products. *Water Sci. Technol.* **26**(3-4), 493-502.

Furumai, H. and Rittmann, B.E. (1994) Interpretation of bacterial activities in nitrification filters by a biofilm model considering the kinetics of soluble microbial products. *Water Sci. Technol.* **30**(11), 147-156.

Gear, C.W. (1971) The automatic integration of ordinary differential equations. *Comm. ACM* **14**, 185-190.

Gonpot, P., Smith, R. and Richter, A. (2000) Diffusion limited biofilm growth. *Model. Sim. Mat. Sci. Eng.* **8**(5), 707-726.

Göpfert, A. and Nehse, R. (1990) *Vektoroptimierung*, Teubner, Leipzig, Germany.

Grady, C.P.L., Daigger, G.T. and Lim, H.C. (1999) *Biological wastewater treatment*, Marcel Dekker, New York, USA.

Gujer, W. (1987) The significance of segregation of biomass in biofilms. *Water Sci. Technol.* **19**, 495-503.

Harremoës, P. (1976) The significance of pore diffusion to filter denitrification. *J. Water Pollut. Control Fed.* **48**(2), 377-388.

Harremoës, P. (1978) Biofilm kinetics. In *Water Pollution Microbiology* (ed. R. Michell), vol. 2, Wiley & Sons, New York, USA.

Harris, N.P. and Hansford, G.S. (1976) A study of substrate removal in a microbial film reactor. *Water Res.* **10**(11), 935-943.

Henze, M., Harremoës, P., Jansen, J.L.C. and Arvin, E. (2002) *Wastewater Treatment*. 3rd ed. Berlin, Springer.

Henze, M., Grady, C.P.L., Gujer, W., Marais, G.v.R. and Matsuo, T. (1987) Activated Sludge Model No. 1, IAWQ, London, UK.

Henze, M., Gujer, W., Mino, T. and Van Loosdrecht, M.C.M. (2000) *Activated sludge models ASM1, ASM2, ASM2d and ASM3* (IWA Task Group on Mathematical Modelling for Design and Operation of Biological Wastewater Treatment; ed. M. Henze, W. Gujer, T. Mino, M.C.M. van Loosdrecht) IWA Scientific & Technical Report, IWA Publishing, London, UK.

Hermanowicz, S.W. (1998) A model of two-dimensional biofilm morphology. *Water Sci. Technol.* **37**(4-5), 219-222.

Hermanowicz, S.W. (2001) A simple 2D biofilm model yields a variety of morphological features. *Math. Biosci.* **169**, 1-14.

Hibbeler, R.C. (1991) *Mechanics of materials*. Macmillan Publ. Comp, NY, USA.

Hinton, E. and Owen, D.R.J. (1977) *Finite element programming*. Academic Press, London, UK.

Horn, H. and Hempel, D.C. (2001) Simulation of substrate conversion and mass transport in biofilm systems. *Chem. Eng. Technol.* **24**(12), A225-A228.

Horn, H., Reiff, H. and Morgenroth, E. (2003) Simulation of growth and detachment in biofilm systems under defined hydrodynamic conditions. *Biotechnol. Bioeng.* **81**(5), 607-617.

Hundsdorfer, W. and Verweer, J. (2003) Numerical Solution of Time-dependent Advection-Diffusion-Reaction Equations, Springer, Berlin.

Hunik, J.H., Bos, C.G., Den Hoogen, M.P., De Gooijer, C.D. and Tramper, J. (1994) Co-immobilized *Nitrosomonas europaea* and *Nitrobacter agilis* cells: validation of a dynamic model for simultaneous substrate conversion and growth in K-carrageenan gel beads. *Biotechnol. Bioeng.* **43**(11), 1153-63.

Hunt, S.M., Hamilton, M.A., Sears, J.T., Harkin, G. and Reno, J. (2003) A computer investigation of chemically mediated detachment in bacterial biofilms. *Microbiology SGM* **149**(5), 1155-1163.

Kissel, J.C., McCarty, P.L. and Street, R.L. (1984) Numerical simulations of mixed-culture biofilms. *J. Environ. Eng.* **110**, 393-411.

Kobayasi, T., Ohmiya, K. and Shimiza, A. (1976) Aproximate expression of effectiveness factor of immobilized enzymes with Michaelis-Menten kinetics. *J. Ferment. Technol.* **54**, 260-263.

Kreft, J.-U., Booth, G. and Wimpenny, J.W.T. (1998) BacSim, a simulator for individual-based modelling of bacterial colony growth. *Microbiology SGM* **144**, 3275-3287.

Kreft, J.-U., Picioreanu, C., Wimpenny, J.W.T. and Van Loosdrecht, M.C.M. (2001) Individual-based modelling of biofilms. *Microbiology SGM* **147**, 2897-2912.

Kreft, J.-U. and Wimpenny, J.W.T. (2001) Effect of EPS on biofilm structure and function as revealed by an individual-based model of biofilm growth. *Water Sci. Technol.* **43**,135-141.

Kugaprasatham, S, Nagaoka, H. and Ohgaki, S. (1992) Effect of turbulence on nitrifying biofilms at non-limiting substrate conditions. *Water Res.* **26**(12), 1629-1638.

LaMotta, E.J. (1976) Internal Diffusion and Reaction in Biological Films. *Environ. Sci. Technol.* **10**(8), 765-769.

Laspidou, C.S. and Rittmann, B.E. (2004) Modeling the development of biofilm density including active bacteria, inert biomass, and extracellular polymeric substances. *Wat. Res.* **38**, 3349-3361.

Laspidou, C.S., Rittmann, B.E. and Karamanos, S.A. (2005) Finite element modeling to expand the UMCCA model to describe biofilm mechanical behavior. In *International Conference Biofilms 2004: Structure and Activity of Biofilms*, Las Vegas, NV, USA.

Lawrence, J.R., Wolfaardt, G.M. and Korber, D.R. (1994) Determination of diffusion coefficients in biofilms by confocal laser microscopy. *Appl. Environ. Microb.* **60**(4), 1166-1173.

Levenspiel, O. (1972) *Chemical reactor Engineering*. Wiley & Sons, New York, USA.

Malmstead, M.J., Brockman, F.J., Valocchi, A.J. and Rittmann, B.E. (1995) Modeling biofilm biodegradation requiring cosubstrates - the quinoline example. *Water Sci. Technol.* **31**(1), 71-84.

Marshall, K.C. (1976) *Interfaces in microbial ecology,* Harvard University Press, Cambridge, Massachusets, USA.

Martins, A.P., Picioreanu, C. and Van Loosdrecht, M.C.M. (2004) Multidimensional dual-morphotype species modelling of activated sludge. *Environ. Sci. Technol.* **38**(21), 5632-5641.

Mehl, M. (2001) Ein interdisziplinärer Ansatz zur dreidimensionalen numerischen Simulation von Strömung, Stofftransport und Wachstum in Biofilmsystemen auf der Mikroskala. Ph.D. thesis, Dept of Computer Science, Munich University of Technology.

Morgenroth, E., van Loosdrecht, M.C.M. and Wanner, O. (2000) Biofilm models for the practitioner. *Water Sci. Technol.* **41**(4-5), 509-512.

Morgenroth, E. and Wilderer, P.A. (2000) Influence of detachment mechanisms on competition in biofilms. *Water Res.* **34**(2), 417-426.

Morgenroth, E. (2003) Detachment - an often overlooked phenomenon in biofilm research and modeling. In *Biofilms in wastewater treatment* (eds., S. Wuertz, P.A. Wilderer and P.L. Bishop), 264-290, IWA Publishing, UK.

Morgenroth, E., Eberl, H.J., Van Loosdrecht, M.C.M., Noguera, D.R., Pizarro, G.E., Picioreanu, C., Rittmann, B.E., Schwarz, A.O. and Wanner, O. (2004) Comparing biofilm models for a single species biofilm system. *Water Sci. Technol.* **49**(11-12), 145-154.

Morton, K.W. (1996) *Numerical solution of convection-Diffusion Problems*, Chapman & Hall, London, UK.

National Research Council (1946) Trickling filters in sewage treatment at military installations. *Sewage Works J.* **18** (5).

National Research Council (2000) Natural Attenuation for Groundwater Remediation. National Academies Press, Washington, DC.

National Research Council (1990) Groundwater Models. Scientific and Regulatory Applications. National Academies Press, Washington, DC.

Nicolella, C., Van Loosdrecht, M.C.M. and Heijnen, J.J. (2000) Wastewater treatment with particulate biofilm reactors. *J. Biotechnol.* **80**(1), 1-33.

Noguera, D.R., Okabe, S. and Picioreanu, C. (1999a) Biofilm modeling: Present status and future directions. *Water Sci. Technol.* **39**(7), 273-278.

Noguera, D.R., Pizarro, G., Stahl, D.A. and Rittmann, B.E. (1999b) Simulation of multispecies biofilm development in three dimensions. *Water Sci. Technol.* **39**(7), 123-130.

Noguera, D.R. and Picioreanu, C. (2004) Results from the Multi-Species Benchmark Problem 3 (BM3) using Two-Dimensional Models. *Water Sci. Technol.* **49**(11-12), 169-176.

Noguera, D.R., Pizarro, G.E. and Regan, J.M. (2004) Modeling Biofilms. In *Microbial Biofilms* (eds. M. Ghannoum and G. A. O'Toole), 222-250, ASM Press, Washington, D.C., USA.

Ohashi, A. and Harada, H. (1994) Adhesion strength of biofilm developed in an attached growth reactor. *Water Sci. Technol.* **29**(10-11), 281-288.

Ohashi, A. and Harada, H. (1996) A novel concept for evaluation of biofilm adhesion strength by applying tensile force and shear force. *Water Sci. Technol.* **34**(5-6), 201-211.

Ohashi, A., Koyama, T., Syutsubo, K. and Harada, H. (1999) A novel method for evaluation of biofilm tensile strength resisting erosion. *Water Sci. Technol.* **39**(7), 261-268.

Okabe, S., Itoh, T., Satoh, H. and Watanabe, Y. (1999) Analyses of spatial distributions of sulfate-reducing bacteria and their activity in aerobic wastewater biofilms. *Appl. Environ. Microbiol.* **65**(11), 5107-16.

Pérez, J., Picioreanu, C. and van Loosdrecht, M.C.M. (2005) Modeling biofilm and floc diffusion processes based on analytical solution of reaction-diffusion equations. *Water Res.* **39**, 1311-1323.

Perry, R.H. and Green, D.W. (1999) *Perry's chemical engineer's handbook*, McGraw-Hill, New York, USA.

Petzold, L. (1983) A description of DASSL: A differential/algebraic system solver. In *Scientific computing* (ed. R. Stepleman), 65-68, IMACS, North Holland.

Peyret, R. and Taylor, T.D. (1990) *Computational Methods for Fluid Flow.* Springer Series in Computational Physics, corrected second printing, Springer, New York, USA.

Picioreanu, C., van Loosdrecht, M.C.M. and Heijnen, J.J. (1998a). Mathematical modeling of biofilm structure with a hybrid differential discrete cellular automaton approach. *Biotechnol. Bioeng.* **58**(1), 101-116.

Picioreanu, C., Van Loosdrecht, M.C.M. and Heijnen, J.J. (1998b) A new combined differential-discrete cellular automaton approach for biofilm modeling: Application for growth in gel beads. *Biotechnol. Bioeng.* **57**(6), 718-731.

Picioreanu, C. (1999) *Multidimensional modeling of biofilm structure.* Ph.D. thesis, Dept of Biotechnology, Delft University of Technology, Delft, The Netherlands.

Picioreanu, C., Van Loosdrecht, M.C.M. and Heijnen, J.J. (1999) Discrete-differential modelling of biofilm structure. *Water Sci. Technol.* **39**(7), 115-122.

Picioreanu, C., Van Loosdrecht, M.C.M. and Heijnen, J.J. (2000a) A theoretical study on the effect of surface roughness on mass transport and transformation in biofilms. *Biotechnol. Bioeng.* **68**, 354-369.

Picioreanu, C., Van Loosdrecht, M.C.M. and Heijnen, J.J. (2000b) Effect of diffusive and convective substrate transport on biofilm structure formation: A two-dimensional modeling study. *Biotechnol. Bioeng.* **69**(5), 504-515.

Picioreanu, C., Van Loosdrecht, M.C.M. and Heijnen, J.J. (2000c) Modelling and predicting biofilm structure. In *Community structure and co-operation in biofilms* (ed. D.G. Allison, P. Gilbert, et al.), 129-166, Cambridge University Press, UK.

Picioreanu, C., van Loosdrecht, M.C.M. and Heijnen, J.J. (2001) Two-dimensional model of biofilm detachment caused by internal stress from liquid flow. *Biotechnol. Bioeng.* **72**(2), 205-218.

Picioreanu, C. and van Loosdrecht, M.C.M. (2002) A mathematical model for initiation of microbiologically influenced corrosion by differential aeration. *J. Electrochem. Soc.* **149**(6), B211-B223.

Picioreanu, C. and Van Loosdrecht, M.C.M. (2003) Use of mathematical modelling to study biofilm development and morphology. In *Biofilms in Medicine, Industry and Environmental Biotechnology - Characteristics, Analysis and Control* (eds. P. Lens, V. O'Flaherty, A.P. Moran, P. Stoodley and T. Mahony), 413-437, IWA Publishing, UK.

Picioreanu, C., Kreft, J.-U., van Loosdrecht, M.C.M. (2004) Particle-based multidimensional multispecies biofilm model. *Appl. Environ. Microb.* **70**(5), 3024-3040.

Picioreanu, C., Batstone, J.D. and Van Loosdrecht, M.C.M. (2005) Multidimensional modelling of anaerobic granules. *Water Sci. Technol.*, in press.

Picioreanu, C., Xavier, J.B. and Van Loosdrecht, M.C.M. (2005) Advances in mathematical modeling of biofilm structure. *Biofilms*, **1**(4), 337-349.

Pizarro, G., Griffeath, D. and Noguera, D.R. (2001) Quantitative cellular automaton model for biofilms. *J. Environ. Eng.* **127**(9), 782-789.

Pizarro, G.E., Teixeira, J., Sepulveda, M. and Noguera, D.R. (2005) Bitwise implementation of a two-dimensional cellular automata biofilm model. *J. Comput. Civil Eng.*, in press.

Ponce Dawson, S., Chen, S. and Doolen, G.D. (1993) Lattice Boltzmann computations for reaction-diffusion equations. *J. Chem. Phys.* **98**(2), 1514-1523.

Press, W.H., Teukolsky, S.A., Vetterling, W.T. and Flannery, B.P. (1997) *Numerical recipes in C: The art of scientific computing*. Cambridge University Press, NY.

Reichert, P. (1994) AQUASIM, a tool for simulation and data analysis of aquatic systems. *Water Sci. Technol.* **30**(2), 21-30.

Reichert, P. and Wanner, O. (1997) Movement of solids in biofilms: significance of liquid phase transport. *Water Sci. Technol.* **36**(1), 321-328.

Reichert, P. (1998a) *AQUASIM 2.0 - User Manual*. Swiss Federal Institute for Environmental Science and Technology (EAWAG), Dübendorf, Switzerland.

Reichert, P. (1998b) *AQUASIM 2.0 - Tutorial*. Swiss Federal Institute for Environmental Science and Technology (EAWAG), Dübendorf, Switzerland.

Rittmann, B.E. and McCarty, P.L. (1980) Model of Steady-State-Biofilm Kinetics. *Biotechnol. Bioeng.* **22**, 2343-2357.

Rittmann, B.E. and McCarty P.L. (1981) Substrate flux into biofilms of any thickness, *J. Environ. Eng.* **108**, 831-849.

Rittmann, B.E. (1982) The effect of shear-stress on biofilm loss rate. *Biotechnol. Bioeng.* **24**(2), 501-506.

Rittmann, B.E. and Brunner, C.W. (1984) The nonsteady-state-biofilm process for advanced organics removal. *J. Water Poll. Control Fed.* **56**(7), 874-880.

Rittmann, B.E. and Manem J. (1992) Development and experimental evaluation of a steady-state, multi-species biofilm model. *Biotechnol. Bioeng.* **39**, 914-922.

Rittmann, B.E., Trinet, F., Amar, D. and Chang, H.T. (1992) Measurement of the activity of a biofilm: Effects of surface loading and detachment on a three-phase, liquid-fluidized-bed reactor. *Wat. Sci. Technol.* **26**(3-4), 585-594.

Rittmann, B.E., Pettis, M., Reeves, H.W. and Stahl, D.A. (1999) How biofilm clusters affect substrate flux and ecological selection. *Water Sci. Technol.* **39**(7), 99-105.

Rittmann, B.E., Schwarz, A.O. and Sáez, P.B. (2000) Biofilms applied to hazardous waste treatment. In *Biofilms II* (ed. J. Bryers), 207-234, John Wiley & Sons, New York, USA.

Rittmann, B.E. and McCarty, P.L. (2001) *Environmental Biotechnology: Principles and Applications*, McGraw Hill, New York, USA.

Rittmann, B.E., Stilwell, D., Garside, J.C., Amy, G.L., Spangenberg, C., Kalinsky, A. and Akiyoshi, E. (2002) Treatment of a colored groundwater by ozone-biofiltration: pilot studies and modeling interpretation. *Water Res.* **36**(13), 3387-3397.

Rittmann, B.E. and Stilwell, D. (2002) Modelling biological processes in water treatment: the integrated biofiltration model. *J Water Supply Res. Technol.-Aqua* **51**(1), 1-14.

Rusten, B., Hellstrom, B.G., Hellstrom, F., Sehested, O., Skjelfoss, E. and Svendsen, B. (2000) Pilot testing and preliminary design of moving bed biofilm reactors for nitrogen removal at the FREVAR wastewater treatment plant. *Water Sci.Technol.* **41** (4-5), 13-20.

Sáez, P.B. and Rittmann, B.E. (1992) Accurate pseudoanalytical solution for steady-state biofilms. *Biotechnol. Bioeng.* **39**, 790-793.

Sethian, J.A. (1996) A fast marching level set method for monotonically advancing fronts. *Proceed. Nat. Acad. Sci. USA* **93**(4), 1591-1595.

Sethian, J.A. (1999) Fast marching methods. *SIAM Review* **41**(2), 199-235.

Silyn-Roberts, G. and Lewis, G. (1997) A technique in confocal laser microscopy for establishing biofilm coverage and thickness. *Water Sci. Technol.* **36**(10), 117-124.

Stadler, W. (1988) Fundamentals of Multi-Critera Optimization. In *Multicriteria Optimisation In Engineering And in Science* (ed. W. Stadler), Plenum Press, New York, USA.

Stahl, D., Flesher, B., Mansfield, H.R. and Montgomery, L. (1988) Use of phylogenetically based hybridization probes for studies of ruminal microbial ecology. *Appl. Environ. Microb.* **54**, 1079-1084.

Stoodley, P., Lewandowski, Z., Boyle, J.D. and Lappin-Scott, H.M. (1998) Oscillation characteristics of biofilm streamers in turbulent flowing water as related to drag and pressure drop. *Biotechnol. Bioeng.* **57**, 536-544.

Stoodley, P., Lewandowski, Z., Boyle, J.D. and Lappin-Scott, H.M. (1999) Structural deformation of bacterial biofilms caused by short-term fluctuations in fluid shear: an in situ investigation of biofilm rheology. *Biotechnol. Bioeng.* **65**, 83-92.

Tchobanoglous, G., Burton, F.L. and Stensel, H.D. (2003) *Wastewater engineering, treatment and reuse*, 4th ed, McGraw Hill, New York.

Van Loosdrecht, M.C.M., Tijhuis, L., Wijdieks, A.M.S. and Heijnen, J.J. (1995a) Biological degradation of organic chemical pollutants in biofilm systems. *Water Sci. Technol.* **31**(1), 163-71.

Van Loosdrecht, M.C.M., Eikelboom, D., Gjaltema, A., Mulder, A., Tijhuis, L. and Heijnen, J.J. (1995b) Biofilm structures. *Water Sci. Technol.* **32**(8), 235-243.

Van Loosdrecht, M.C.M., Picioreanu, C. and Heijnen, J.J. (1997) A more unifying hypothesis for biofilm structures. *FEMS Microbiol. Ecol.* **24**(2), 181-183.

Wanner, O. and Gujer, W. (1984) Competition in biofilms. *Wat. Sci. Technol.* **17**(2/3), 27-44.

Wanner, O. and Gujer, W. (1986) A multispecies biofilm model. *Biotechnol. Bioeng.* **28**, 314-328.

Wanner, O. (1989) Modeling population dynamics. In *Structure and Function of Biofilms* (eds. W.G. Characklis and P.A. Wilderer), 91-110, John Wiley & Sons, New York, USA.

Wanner, O. and Reichert, P. (1996) Mathematical modeling of mixed-culture biofilms. *Biotechnol. Bioeng.* **49**(2), 172-184.

Wanner, O. (2002) Modeling of Biofilms. In *Encyclopedia of Environmental Microbiology* (ed. G. Bitton), 2083-2094, John Wiley & Sons, New York, USA.

Watanabe, Y., Okabe, S., Hirata, K. and Masuda, S. (1995) Simultaneous removal of organic materials and nitrogen by micro-aerobic biofilms. *Wat. Sci. Technol.* **31**(1), 170-177.

Wilderer, P.A. and Characklis, W.G. (1989) Structure and function of biofilms. In *Structure and function of biofilms* (eds. W.G. Characklis and P.A. Wilderer), 5-17, John Wiley & Sons, New York, USA.

Williamson, K. and McCarty, P.L. (1976) A model of substrate utilization by bacterial films. *J. Water Pollut. Control Fed.* **48** (1), 9-24.

Wilson, E.J. and Geankoplis, C.J. (1966) Liquid mass transfer at very low Reynolds numbers in packed beds. *Ind. Eng. Chem. Fundam.* **5**, 9-14.

Wimpenny, J.W.T. and Colasanti, R. (1997) A unifying hypothesis for the structure of microbial biofilms based on cellular automaton models. *FEMS Microbiol. Ecol.* **22**, 1-16.

Wood, B.D. and Whitaker, S. (1998) Diffusion and reaction in biofilms. *Chem. Eng. Sci.* **53**(3), 397-425.

Wood, B.D. and Whitaker, S. (1999) Cellular growth in biofilms. *Biotechnol. Bioeng.* **64**, 656-670.

Xavier, J.B., White, D.C. and Almeida, J.S. (2003) Automated biofilm morphology quantification from confocal laser scanning microscopy imaging. *Water Sci. Technol.* **47**(5), 31-37.

Xavier, J.B., Picioreanu, C. and Van Loosdrecht, M.C.M. (2004) Assessment of three-dimensional biofilm models through direct comparison with confocal microscopy imaging. *Water Sci. Technol.* **49**(1), 177-185.

Xavier, J.B., Picioreanu, C. and van Loosdrecht, M.C.M. (2005a) A Framework for Multidimensional Modelling of Activity and Structure of Multispecies Biofilms. *Environ. Microbiol.*, Early View online.

Xavier, J.B., Picioreanu, C. and Van Loosdrecht, M.C.M. (2005b) A general description of detachment for multidimensional modelling of biofilms. *Biotechnol. Bioeng.*, Early View online.

Xavier, J.B., Picioreanu, C. and Van Loosdrecht, M.C.M. (2005c) A modelling study of the activity and structure of biofilms in biological reactors. *Biofilms* 1(4), 377-391.

Yamane, T. (1981) On approximate expression of effectiveness factor of immobilized biocatalysts. *J. Ferment. Technol.* **59**, 375-381.

Zhang, T.C. and Bishop, P.L. (1994a) Structure, activity and composition of biofilms. *Water Sci. Technol.* **29**(7), 335-344.

Zhang, T.C. and Bishop, P.L. (1994b) Density, Porosity, and Pore Structure of Biofilms. *Water Res.* **28**(11), 2267-2277.

Zhang, T.C. and Bishop, P.L. (1994c) Evaluation of tortuosity factors and effective diffusivities in biofilms. *Wat. Res.* **28**(11), 2279-2287.

Zobell, C.E. and Anderson, D.Q. (1936) Observations on the multiplication of bacteria in different volumes of stored seawater and the influence of oxygen tension and solid surfaces. *Biol. Bull. Woods Hole* **71**, 342.

Index

Printed in the United Kingdom
by Lightning Source UK Ltd.
110437UKS00002B/3-74

9 781843 390879